普通高等教育“十三五”规划教材

云计算平台搭建项目化教程

符龙生　王　岩　主　编
颜汝南　李　健　副主编
黄　斌　梁其钰
王　忠　主　审

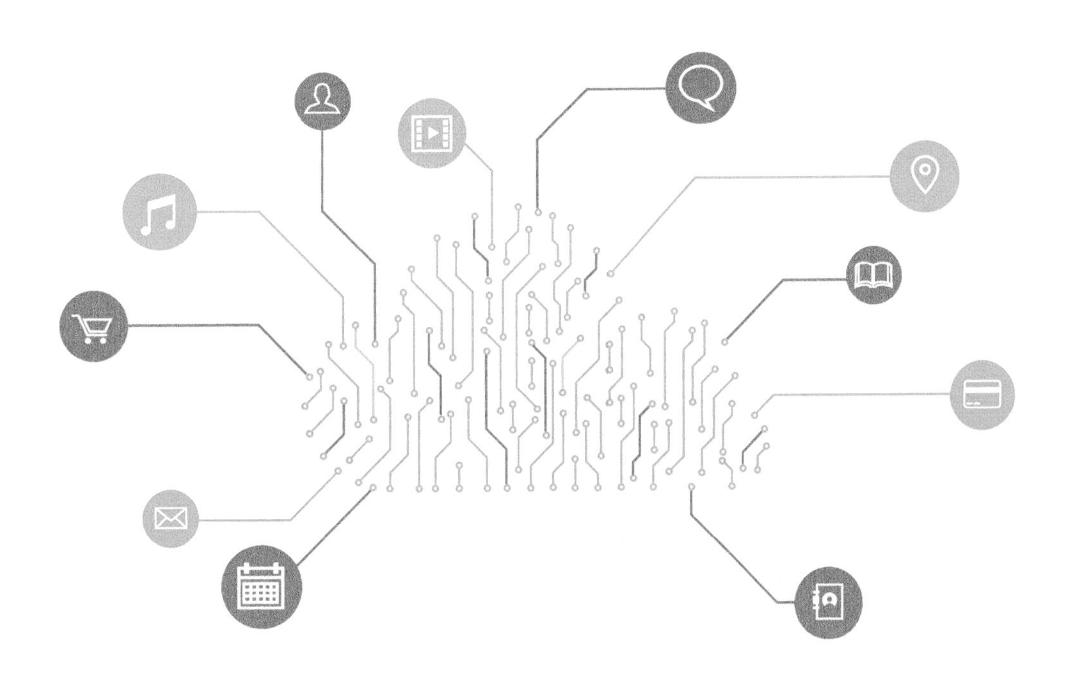

中国铁道出版社有限公司
CHINA RAILWAY PUBLISHING HOUSE CO., LTD.

内容简介

本书基于开源的 OpenStack 云计算架构,以企业对云平台应用的需求为主线搭建私有云,主要包括认识云计算及搭建云计算基础服务平台、认证服务、消息队列服务、镜像服务、网络服务、计算服务、存储服务和高级控制服务等相关内容,最后设置了一个综合实训,帮助读者梳理云计算基础架构平台的部署、配置和管理工作。

本书适合作为普通高等院校计算机网络专业或大数据专业的教材,也可作为企业网络、大数据等专业的培训教材,以及计算机爱好者的自学用书。

图书在版编目（CIP）数据

云计算平台搭建项目化教程/符龙生,王岩主编.—北京:
中国铁道出版社有限公司,2020.2（2024.8 重印）
普通高等教育“十三五”规划教材
ISBN 978-7-113-26605-9

Ⅰ.①云… Ⅱ.①符… ②王… Ⅲ.①云计算-高等学校-教材 Ⅳ.①TP393.027

中国版本图书馆 CIP 数据核字(2020)第 021402 号

书　　名：云计算平台搭建项目化教程
作　　者：符龙生　王　岩

策　　划：唐　旭　　编辑部电话：(010)63549508
责任编辑：陆慧萍　卢　笛
封面设计：尚明龙
责任校对：张玉华
责任印制：樊启鹏

出版发行：中国铁道出版社有限公司（100054，北京市西城区右安门西街 8 号）
网　　址：https://www.tdpress.com/51eds/
印　　刷：北京九州迅驰传媒文化有限公司
版　　次：2020 年 2 月第 1 版　2024 年 8 月第 5 次印刷
开　　本：787 mm×1 092 mm 1/16　印张：11　字数：245 千
书　　号：ISBN 978-7-113-26605-9
定　　价：36.80 元

前　言

PREFACE

计算机技术经历了从大型主机、个人计算机、客户/服务器计算模式到今天的互联网计算模式的演变，尤其是互联网 Web 2.0 技术的应用，使计算能力需求更多地依赖于通过互联网连接的远程服务器资源。作为资源的提供者，需要具备超高的计算性能、海量的数据存储能力、网络通信能力和随时的扩展能力。在多种应用需求的推动下催生了虚拟化技术和云计算技术。当今，云计算平台已经成为云存储技术、大数据分析、“互联网 +”技术等信息技术应用服务基础平台，在信息技术的发展过程中起着支撑作用。云计算是推动信息技术能力实现按需供给，促进信息技术和数据资源充分利用的全新业态，是信息化发展的重大变革和必然趋势。发展云计算，有利于分享信息知识和创新资源，降低全社会创业成本，培育形成新产业和新消费热点，对稳增长、惠民生和建设创新型国家具有重要意义。

本书为教育部首批“新工科”研究与实践项目——面向新经济的计算机网络技术专业改造升级路径探索与实践（项目编号:204）课题专项资助成果。本书是校企合作产教融合后的实践产物，基于开源的 OpenStack 云计算架构，解决高等院校云计算专业或相关专业的云计算架构搭建与应用的教学需求。遵循以项目为驱动、任务为目标的编写思路，每个项目分为若干个子任务，每个任务首先提出具体的任务要求，然后介绍任务的相关知识，最后介绍完成任务的具体操作步骤，做到基础知识介绍具有针对性，任务目标操作实现具体化。

本书涵盖的内容

全书共分 9 个项目，下面分别对每个项目进行简单的介绍：

（1）项目一认识云计算及搭建云计算基础服务平台，介绍了企业对云平台应用的需求，云平台系统架构设计，云平台基础部署工作以及验证安装基础工作。

（2）项目二认证服务，介绍配置 Keystone 应用环境、管理认证用户及创建租户、创建用户账号和绑定用户权限。

（3）项目三消息队列服务，主要介绍消息队列以及消息队列的基本操作。

（4）项目四镜像服务，主要介绍镜像服务基本操作、制作 CentOS 7.2 镜像及镜像上传。

（5）项目五网络服务，主要介绍网络服务的基础操作，创建各部门网络子网以及创建外来访问使用的网络。

（6）项目六计算服务，主要介绍 Nova 及其底层的运行机制、原理以及正常启动、关闭和重建虚拟机等操作。

(7)项目七存储服务,主要介绍块存储组件 Cinder 的基本服务,对 Cinder 组件后端逻辑卷进行扩容,通过 Cinder 组件的 CLI 命令行和通过 Dashboard 界面完成块存储任务;对象存储组件 Swift 的基本操作及其应用。

(8)项目八高级控制服务,主要介绍编配服务(Heat 编排),完成编配服务任务和监控服务的基本操作以及进行数据查看及数据库备份的方法。

(9)项目九综合实训,主要训练学生综合运用前面所学云技术的能力,提高学生的综合技术运用能力。

阅读本书需要的准备

本书中涉及的硬件设备推荐使用带有双网卡(或多网卡)的两台高性能服务器。两台服务器 CPU 内核数在 4 个以上,控制节点内存需大于 2 GB,计算节点内存需大于 4 GB,两个节点的存储空间都需在 100 GB 以上。如果上述的实验环境都不具备,可以在高性能计算机上安装 VMware Workstation 12 或者 15 以上的版本,通过 VMware 中的虚拟机来模拟控制节点和计算节点,这样的操作环境对计算机硬件要求较高,通常计算机配备四核 4 线程以上的 CPU、8 GB 以上的内存以及 300 GB 以上的空余存储空间。推荐配置为 I 7 处理器,16 GB 内存,500 GB 空余磁盘空间。本书中任务实施所用的两个软件包是 CentOS-7-x86_64-DVD-1511. iso 的 iso 镜像文件为 CentOS 7 版本的安装包,在控制节点和计算节点上均使用 CentOS 7.2 最小化安装作为先电 IaaS 的底层的操作系统。Xiandian-IaaS-2.2. iso 为基于 OpenStack 的先电云平台软件包,该软件包用于构建本地 YUM 源并安装 IaaS 云平台的各个组件。编者的电子邮箱为 153170997@qq. com,如读者需要本书相关的软件资源,可以与编者或出版社联系。

本书由符龙生、王岩任主编,颜汝南、李健、黄斌、梁其钰任副主编。具体分工如下:项目一由梁其钰编写,项目二至四由符龙生编写,项目五由颜汝南编写,项目六由黄斌编写,项目七由李健编写,项目八、九由王岩编写。全书由王忠主审。

在本书的编写过程中,参考了大量的 OpenStack 的技术资料、全国职业技能大赛资料,同时汲取了网络上许多计算机爱好者或同仁的宝贵经验,在此表示诚挚的谢意。

由于编者水平有限,书中的不妥和疏漏在所难免,恳请各位专家和广大读者及时批评指正。

编　者

2019 年 8 月

CONTENTS

目 录

项目一 认识云计算及搭建云计算基础服务平台

项目综述

小张刚从学校毕业，被某公司聘用为云计算助理工程师，公司现准备将原有的计算机服务器改造成云计算服务平台。为此小张必须了解云计算的基础概念及搭建云计算平台的相关知识，以便提出详细的改建方案和实施步骤。接下来需要对公司的应用需求进行调研，在此基础上要进行公司云计算平台的系统环境设计和系统搭建的基础安装工作。为此，小张当前要完成的任务如下：

视频

项目一综述

- 认识云计算。
- 公司云平台系统架构的设计。
- 云平台系统安装基础工作。

项目目标

【知识目标】

- 了解云计算的起源及发展历程。
- 理解云计算相关概念。
- 了解 OpenStack 项目。

【技能目标】

- 了解准备 OpenStack 搭建云计算平台项目所需的软件资源包。
- 设置各节点的名称、IP 地址和安装各节点的操作系统。
- 掌握各节点安装云平台的基础工作。

【职业能力】

根据企业的应用需求，使用 OpenStack 开源软件，搭建适合企业的云平台。

任务一　认识云计算

任务要求

小张有必要了解云计算的基础概念及搭建云计算平台的相关知识，以便提出详细的改建方案和实施步骤。

核心概念

云计算平台，又称云平台，是指基于硬件资源和软件资源的服务，提供计算、网络和存储能力。云计算平台可以划分为3类：以数据存储为主的存储型云平台，以数据处理为主的计算型云平台以及计算和数据存储处理兼顾的综合云计算平台。

知识准备

云计算的起源及发展历程

云计算的起源及发展历程：

1951年，UNIVAC－1诞生，这是世界上第一台商用计算机系统，被用来进行美国人口普查，正式标志着计算机进入商业应用时代。

1959年，英国计算机科学家Christopher Strachey发表关于虚拟化论文，其虚拟化理论是如今云计算基础架构的基础理论之一。

1961年，计算机科学家John McCarthy发表公开演说：“如果计算机在未来流行开来，那么未来计算机也可以像电话一样成为公用设施……计算机应用也将成为一种全新的、重要的产业基石。”

1969年，ARPANET项目的首席科学家Leonard Kleinrock表示：“计算机网络现在还处于初期阶段，但随着网络的进步和复杂化，未来可能看到‘计算机应用’的扩展……”

1984年，Sun公司（已于2009年被Oracle公司收购）的联合创始人John Gage提出“网络就是计算机”的猜想，用于描述分布式计算技术带来的新世界。今天的云计算发展也证实了这一猜想，并逐步地将这一理念变成现实。

1996年，网格计算Globus开源网格平台起步，网格技术也被普遍认为是云计算技术的前身技术之一。

1998年，VMware（威睿公司）成立并首次引入X86的虚拟技术。

1999年，MarcAndreessen创建LoudCloud，是世界上第一个商业化的IaaS平台。

2004年，Web 2.0会议举行，Web 2.0成为技术流行词，互联网发展进入新阶段。

2006年，“云计算”这一术语正式出现在商业领域，Google的CEO在搜索引擎大会上提出云计算，Amazon推出其弹性计算云（EC2）服务。

2008年，中国第一个获得自主知识产权的基础架构云（IaaS）产品BingoCloudOS（品高云）发行1.0版。

2009年，NIST（National Institute of Standards and Technology，美国国家标准与技术研究院）发布了被业界广泛接受的云计算定义“一种标准化的IT性能（服务、软件或者基础设施），以按使用付费和自助服务方式，通过Internet技术交付”。

云计算成为IT领域最引人瞩目的热点之一，也是当前企业IT建设正在投入或者将要投入的重要领域之一。因此，云计算的发展和企业IT建设也在相辅相成地进行着。

1. 第一时期：集群化时期

第一时期，是企业IT建设的初步启蒙阶段。在这一阶段中，企业将自己的IT

硬件资源进行物理型集中，将分散的资源组合成规模化的数据中心，为企业统一提供集中式的基础设施服务。在这一过程中，企业不断地实施业务与数据整合，大多数企业的数据中心开始完成自身标准化，使得已有业务可扩展性和新业务能够不断规划。同时，这样的集中化时期也产生了不同的问题，譬如数据传输中断、数据中心业务崩塌带来的数据中断问题，因此企业在建设的后期也开始注重数据中心容灾建设，特别是一些以数据为中心的热点行业，如金融、网络等行业开展了容灾建设热潮，且大多数企业的容灾建设级别非常高，基本都处于面向应用级容灾（零丢失、零中断），希望业务能在不间断的环境下运行。

2. 第二时期：网格计算时期

第二时期，是网格计算时期。计算网格为计算资源提供了一个平台，使其能组织成一个或多个逻辑池。这些逻辑池统一协调为一个高性能分布式系统，又被称为超级虚拟计算机。网格计算与集群化的区别在于，网格系统更加松耦合，也更加分散，因此网格系统可以包含异构、异地的不同技术资源。从 20 世纪 90 年代开始，网格技术就被作为计算科学的一部分，它的一些通用特效譬如可恢复性、可扩展性、网络接入和资源池等属性仍然影响云计算的方方面面。网格计算以中间件层为基础，且中间件层在计算资源上部署的。这些 IT 资源构成一个网格池，实现统一的分配和协调，现在也普遍认为网格计算这一时期是云计算发展的雏形时期。

3. 第三时期：传统虚拟化时期

第三时期，是企业 IT 建设的蓬勃发展时期。这一阶段数据中心 IT 设备飞速扩张，但是随之而来的就是系统建设成本高、开发建设时期长等问题，严重滞后企业 IT 建设发展速度。虽然企业内部进行了 IT 建设业务模板标准化，但这样的情形下，哪怕是复制型的业务建设都无法降低软硬件采购成本、调试运行成本和缩短业务开发周期时长。因为标准化只是小范围的规划，集中型的大规模 IT 设施带来了大量系统利用率不足的问题，不同的系统独占了各类硬件资源，造成极大的数据中心资源浪费。所以这一时期，最迫切需要解决的问题就是成本过高、IT 运行灵活性较低、资源利用率低下。虚拟化技术在这一时期被广泛利用，它是一个技术平台，用于创建 IT 资源的虚拟实例。虚拟化软件层允许物理 IT 资源提供自身的多个虚拟映像，这样一来，多个用户就可以共享底层处理能力。

虚拟化将多台服务器整合，屏蔽了底层设备的差异之处转向统一提供处理能力，实现了物理服务器资源利用率的提升。在虚拟化技术出现之前，软件只能被绑定在静态硬件环境中，而虚拟化则打断了这种软硬件之间的依赖性，因为虚拟化环境可以模拟对硬件的需求，从而使得企业数据中心服务器可以大幅度提升计算效能，解决 IT 设施的能耗与空间问题。

4. 第四时期：云计算时期

这一时期也正是当前所处的时期。对企业来说，投入一大笔费用（软硬件成本）建设的数据中心，后续更新迭代也是一笔不小的开支。硬件设备在数据中心建立 5 年左右就面临着老化、陈旧等问题，软件技术也有着不断升级的压力。另一方面，IT 的投入与业务发展十分相关，即使在虚拟化后，也难以解决不断增加的业务对自由的变化需求。此时企业希望 IT 资源能够弹性扩展、按需服务，将服务作为

IT 的核心，其主要的商业驱动力是容量规划、成本需求和组织灵活性。云计算架构可以由企业自己构建，也可以采用第三方云设施，基本趋势就是企业采用租用 IT 资源的方式来实现业务需求，如同获取水力、电力一样，无须知道如何建设但可以方便使用。

5. 云计算的未来

云计算现阶段已经成为 IT 企业运行的标配。2016 年 7 月 15 日，中国银监会发布了《中国银行业信息科技"十三五"发展规划监管指导意见（征求意见稿）》，其中指出银行业应稳步实施架构迁移，到"十三五"末期，面向互联网场景的重要信息系统全部迁移至云计算架构平台，其他系统迁移比例不低于 60%。不难看出，银监会发布了关于金融云的使用指导，要求五年内 60% 的金融业务要上云。这就是全行业的一个缩影，传统行业对于云计算需求的趋势无可阻挡，可以肯定，未来云计算的重点将在于各行业的业务上，云计算将大面积地用于金融、教育、军工等各行各业。

云计算，从一开始的"要不要用"到"什么时候用"再到"必须要用"，云计算已逐步成为企业 IT 建设的标配，真正地拥抱云计算的时代已经到来。

视频

云计算的基本概念

1. 了解云计算的基本概念

1）云计算的定义

云计算到底是什么呢？在这个问题上，可谓众说纷纭。比如，在维基百科上的定义是"云计算是一种基于互联网的计算新方式，通过互联网上异构、自治的服务为个人和企业用户提供按需即取的计算"；著名咨询机构 Gartner 将云计算定义为"云计算是利用互联网技术来将庞大且可伸缩的 IT 能力集合起来作为服务提供给多个客户的技术"；而 IBM 则认为"云计算是一种新兴的 IT 服务交付方式，应用、数据和计算资源能够通过网络作为标准服务在灵活的价格下快速地提供最终用户"。

虽然这几个定义都有一定的道理，但还没抓住云计算的核心。云计算的模式如图 1-1所示，云计算是新一代 IT 模式，它能在后端庞大的云计算中心的支撑下为用户提供更方便的体验和更低廉的成本。

图 1-1　云计算的模式

具体而言，由于在后端有规模庞大、高度自动化和高可靠性的云计算中心的存在，人们只要接入互联网，就能非常方便地访问各种基于云的应用和信息，并免去了安装和维护等烦琐操作，同时，企业和个人也能以低廉的价格来使用这些由云计算中心提供的服务或者在云中直接搭建其所需的信息服务。在收费模式上，云计算和水电等公用事业非常类似，用户只需为其所使用的部分付费。对云计算的使用者

（主要是个人用户和企业）来讲，云计算将会在用户体验和成本这两方面给他们带来很多非常实在的好处。

2）云计算的特征

①超大规模：大多数云计算中心都具有相当的规模。比如，Google 云计算中心已经拥有几百万台服务器，而 Amazon 、IBM 、微软、Yahoo 等企业所掌控的云计算规模也毫不逊色，并且云计算中心能通过整合和管理这些数目庞大的计算机集群来赋予用户前所未有的计算和存储能力。

②抽象化：云计算支持用户在任意位置、使用各种终端获取应用服务，所请求的资源都来自“云”，而不是固定的有形的实体。应用在“云”中某处运行，但实际上用户无须了解、也不用担心应用运行的具体位置，这样能有效地简化应用的使用。

③高可靠性：在这方面，云计算中心在软硬件层面采用了诸如数据多副本容错、心跳检测和计算节点同构可互换等措施来保障服务的高可靠性，还在设施层面上的能源、制冷和网络连接等方面采用了冗余设计来进一步确保服务的可靠性。

④通用性：云计算中心很少为特定的应用存在，但其有效支持业界大多数的主流应用，并且一个“云”可以支撑多个不同类型应用的同时运行，并保证这些服务的运行质量。

⑤高可扩展性：用户所使用“云”的资源可以根据其应用的需要进行调整和动态伸缩，并且再加上前面所提到的云计算中心本身的超大规模，使得“云”能有效地满足应用和用户大规模增长的需要。

⑥按需服务：“云”是一个庞大的资源池，用户可以按需购买，就像自来水、电和煤气等公用事业那样根据用户的使用量计费，并无须任何软硬件和设施等方面的前期投入。

⑦廉价：首先，由于云计算中心本身巨大规模所带来的经济性和资源利用率的提升；其次，“云”大都采用廉价和通用的 X86 节点来构建，因此用户可以充分享受云计算所带来的低成本优势，通常只要花费几百美元就能完成以前需要数万美元才能完成的任务。

⑧自动化：云中不论是应用、服务和资源的部署，还是软硬件的管理，都主要通过自动化的方式来执行和管理，从而极大地降低整个云计算中心庞大的人力成本。

⑨节能环保：云计算技术能将许许多多分散在低利用率服务器上的工作负载整合到云中，来提升资源的使用效率，而且云由专业管理团队运维，所以其 PUE（Power Usage Effectiveness ，电源使用效率值）值和普通企业的数据中心相比出色很多。比如，Google 数据中心的 PUE 值在 1.2 左右，也就是说，每一美元的电力花在计算资源上，只需再花两美分电力在制冷等设备，而常见的 PUE 在 2 和 3 之间，并且还能将云建设在水电厂等洁净资源旁边，这样既能进一步节省能源方面的开支，又能保护环境。

⑩完善的运维机制：在“云”的另一端，有全世界最专业的团队来帮用户管理信息，有全世界最先进的数据中心来帮用户保存数据。同时，严格的权限管理策略可以保证这些数据的安全。这样，用户无须花费重金就可以享受到最专业的服务。

由于这些特征的存在，使得云计算为用户提供更方便的体验和更低廉的成本，同时这些特点也是为什么云计算能脱颖而出，并且被大多数业界人员所推崇的原因之一。

3）云计算的服务架构

虽然云计算涉及很多产品与技术，表面上看起来的确有点纷繁复杂，但是云计算本身还是有迹可循和有理可依的，图 1－2就是一套云计算的服务架构示意图。

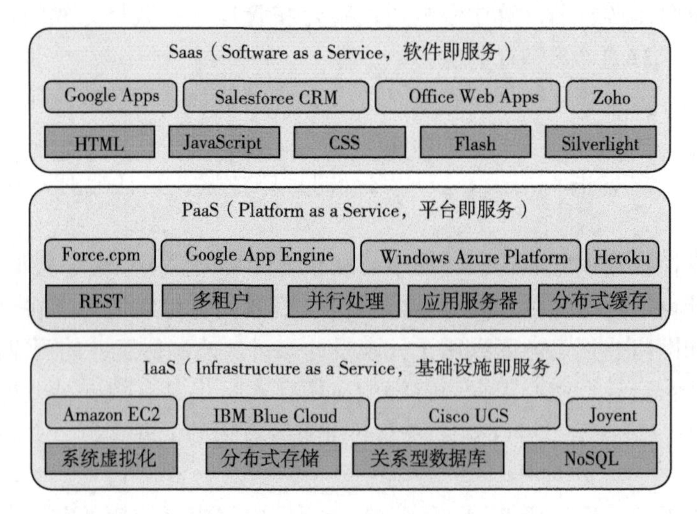

图 1－2　云计算的服务架构

云架构分为服务和管理两大部分。

在服务方面，主要以提供用户基于云的各种服务为主，共包含三个层次：其一是 SaaS（Software as a Service，软件即服务），这层的作用是将应用主要以基于 Web 的方式提供给客户；其二是 PaaS（Platform as a Service，平台即服务），这层的作用是将一个应用的开发和部署平台作为服务提供给用户；其三是 IaaS（Infrastructure as a Service，基础架构即服务），这层的作用是将各种底层的计算（如虚拟机）和存储等资源作为服务提供给用户。从用户角度而言，这三层服务之间的关系是独立的，因为它们提供的服务是完全不同的，而且面对的用户也不尽相同。但从技术角度而言，云服务这三层之间的关系并不是独立的，而是有一定依赖关系的，如一个 SaaS 层的产品和服务不仅需要使用到 SaaS 层本身的技术，而且还依赖 PaaS 层所提供的开发和部署平台或者直接部署于 IaaS 层所提供的计算资源上。还有，PaaS 层的产品和服务很有可能构建于 IaaS 层服务之上。

4）云计算的 4 种模式

虽然从技术或者架构角度看，云计算都是比较单一的，但是在实际情况下，为了适应用户不同的需求，它会演变为不同的模式。在 NIST 的名为 *The NIST Definition of Cloud Computing* 的这篇关于云计算概念的著名文章中，共定义了云的 4 种模式，分别是：公有云、私有云、混合云和行业云。接下来，将详细介绍每种模式的概念、构建方式、优势、不足之处及其对未来的展望等。

①公有云。公有云是当前主流的也是最受欢迎的云计算模式。它是一种对公众开放的云服务，能支持数目庞大的请求，而且因为规模的优势，其成本偏低。公有

云由云供应商运行，为最终用户提供各种各样的 IT 资源。云供应商负责从应用程序、软件运行环境到物理基础设施等 IT 资源的安全、管理、部署和维护。在使用 IT 资源时，用户只需为其所使用的资源付费，无须任何前期投入，所以非常经济。在公有云中，用户不清楚与其共享和使用资源的还有其他哪些用户，整个平台是如何实现的，甚至无法控制实际的物理设施，但是云服务提供商须保证其所提供的资源具备安全和可靠等非功能性需求。

许多 IT 巨头都推出了它们自己的公有云服务，包括 Amazon 的 AWS、微软的 Windows Azure Platform、Google 的 Google Apps 与 Google App Engine 等。一些过去著名的 VPS 和 IDC 厂商也推出了它们自己的公有云服务，如 Rackspace 的 Rackspace Cloud 和国内世纪互联的 CloudEx 云快线等。

②私有云。关于云计算，虽然人们谈论最多的莫过于以 Amazon EC2 和 Google App Engine 为代表的公有云，但是对许多大中型企业而言，因为很多限制和条款，它们在短时间内很难大规模地采用公有云技术，可是它们也期盼云所带来的便利，所以引出了私有云这一云计算模式。私有云主要为企业内部提供云服务，不对公众开放，在企业的防火墙内工作，并且企业 IT 人员能对其数据、安全性和服务质量进行有效控制。与传统的企业数据中心相比，私有云可以支持动态灵活的基础设施，降低 IT 架构的复杂度，使各种 IT 资源得以整合和标准化。

在私有云界，主要有两大联盟：其一是 IBM 与其合作伙伴，主要推广的解决方案有 IBM Blue Cloud 和 IBM CloudBurst；其二是由 VMware、Cisco 和 EMC 组成的 VCE 联盟，它们主推的是 Cisco UCS 和 vBlock。在实际案例方面，已经建设成功的私有云有采用 IBM Blue Cloud 技术的中化云计算中心和采用 Cisco UCS 技术的 Tutor Perini 云计算中心。

③混合云。混合云虽然不如前面的公有云和私有云常用，但已经有类似的产品和服务出现。顾名思义，混合云是把公有云和私有云结合到一起的方式，即它是让用户在私有云的私密性和公有云灵活的低廉之间做一定权衡的模式。比如，企业可以将非关键的应用部署到公有云上来降低成本，而将安全性要求很高、非常关键的核心应用部署到完全私密的私有云上。

现在混合云的案例非常少，最相关的就是 Amazon VPC（Virtual Private Cloud，虚拟私有云）和 VMware vCloud 了。比如，通过 Amazon VPC 服务能将 Amazon EC2 的部分计算能力接入到企业的防火墙内。

④行业云。行业云虽然较少提及，但是有一定的潜力，主要指的是专门为某个行业的业务设计的云，并且开放给多个同属于这个行业的企业。

虽然行业云还没有一个成熟的案例，但盛大的开放平台颇具行业云的潜质，因为它能将其整个云平台共享给多个小型游戏开发团队，这样这些小型团队只需负责游戏的创意和开发即可，其他和游戏相关的烦琐的运维可转交给盛大的开放平台来负责。

2. 了解 OpenStack 项目

OpenStack 是一个由 NASA（美国国家航空航天局）和 Rackspace 合作研发并发起的，以 Apache 许可证授权的自由软件和开放源代码项目。

OpenStack 是一个开源的云计算管理平台项目，由几个主要的组件组合起来完成具体工作。OpenStack 支持绝大多数类型的云环境，项目目标是提供实施简单、可大规模扩展、丰富、标准统一的云计算管理平台。OpenStack 通过各种互补的服务提供了基础设施即服务（IaaS）的解决方案，每个服务提供 API 以进行集成。

OpenStack 是一个旨在为公共及私有云的建设与管理提供软件的开源项目。它的社区拥有超过 130 家企业及 1 350 位开发者，这些机构与个人都将 OpenStack 作为基础设施即服务（IaaS）资源的通用前端。OpenStack 项目的首要任务是简化云的部署过程并为其带来良好的可扩展性。本文希望通过提供必要的指导信息，帮助大家利用 OpenStack 前端来设置及管理自己的公共云或私有云。

OpenStack 云计算平台，帮助服务商和企业内部实现类似于 Amazon EC2 和 S3 的云基础架构服务。OpenStack 包含两个主要模块：Nova 和 Swift，前者是 NASA 开发的虚拟服务器部署和业务计算模块；后者是 Rackspace 开发的分布式云存储模块，两者可以一起用，也可以分开单独用。OpenStack 除了有 Rackspace 和 NASA 的大力支持外，还有包括 Dell、Citrix、Cisco、Canonical 等重量级公司的贡献和支持，发展速度非常快，有取代另一个业界领先开源云平台 Eucalyptus 的态势。

OpenStack 是一个综合的云计算管理平台，在 OpenStack 项目中包含各种各样的组件，如提供身份验证的 Keystone 组件、提供计算服务的 Nova 组件、提供镜像服务的 Glance 组件、提供对象存储的 Swift 组件、提供网络服务的 Neutron 组件、提供块存储服务的 Cinder 组件、提供面板服务的 Horizon 组件等，如图 1-3 所示。

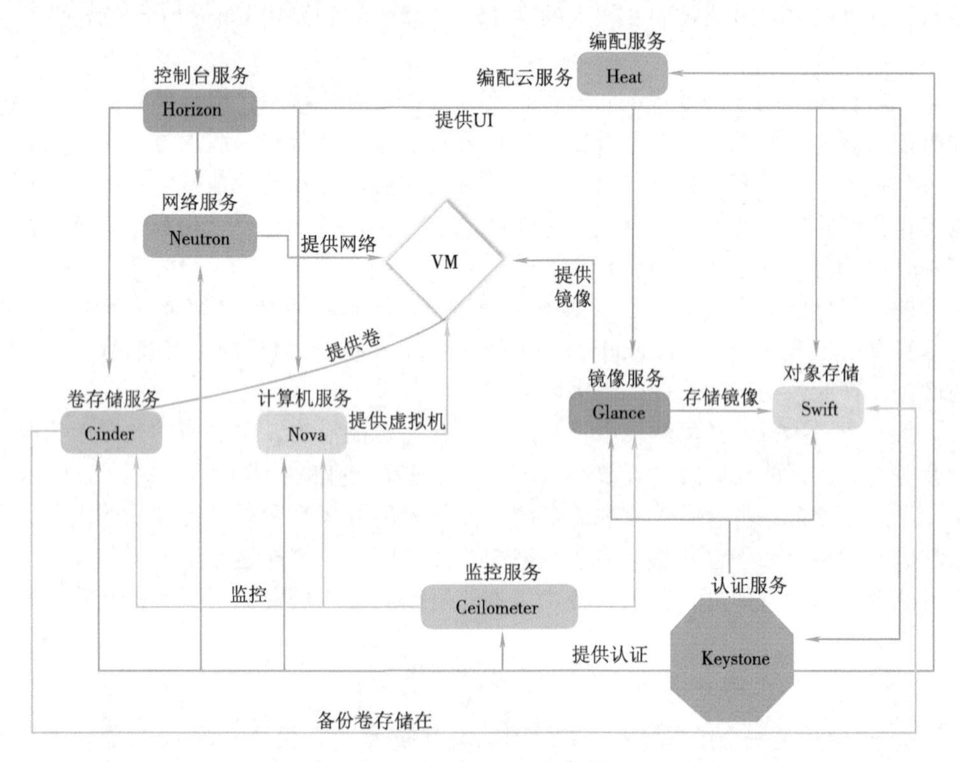

图 1-3　OpenStack 框架图

提供认证服务的 Keystone 组件在 OpenStack 中主要负责用户、租户、服务和服务端点的管理，同时也是 OpenStack 中各个组件间认证的核心，通过了服务的注册功能，并且可以支持 SQL、PAM、LDAP 等后端的认证。

提供计算服务的 Nova 组件主要负责虚拟机实例的调度分配以及实例的创建、启停、迁移、重启等操作，从而管理云中实例的生命周期，是整个 OpenStack 云中的组织控制器。运行的虚拟实例支持的 Hypervisor 有 KVM 等，默认的是 KVM，同时提供了与 Amazon 兼容的 API 接口。

提供镜像服务的是 Glance 组件，能够实现镜像的创建、镜像的快照管理、镜像模板等，同时支持各种镜像，如无格式的 raw、默认格式 qcow、VMware 格式的 vhd 等，而实际的镜像文件通常是不保存在本地的，而是保存在存储的后端，如 Swift、文件系统、Amazon S3 等。镜像格式：raw、qcow、vhd、vmdk、iso；后端存储：Swift、Filesystem、Amazon S3。

提供对象存储的是 Swift 组件，主要提供了存取数据的应用服务，与 Amazon S3 比较类似，通常用于保存非结构化数据，如通常作为 Glance 组件镜像的存储后端，或作为云盘等应用。

提供网络服务的是 Neutron 组件，是基于软件定义网络的思想，实现网络资源的软件化管理，支持各种类型的插件，实现多租户网络的隔离，也可以与支持硬件和软件的网络解决方案进行集成。

提供块存储的是 Cinder 组件，为虚拟化实例提供卷的持久化存储服务，同时支持对卷的快照、备份等管理，和 Amazon EBS 组件较为相似，基于插件的架构，便于扩展 。

提供 Web 统一化管理界面的是 Horizon 组件，主要提供了自动化仪表板的管理服务，实现对用户、租户、卷、网络等绝大多数资源的图形化管理。

提供监控和计量服务的是 Ceilometer 组件，主要实现对 OpenStack 平台中各个组件信息的统一化采集处理，从而实现对 OpenStack 各个组件状态的监控以及流量统计等功能。

（1）了解 OpenStack 的技术资源如下：

①OpenStack 官网：

· http:/www. OpenStack. org/。

· http://wiki. OpenStack. org/wiki/Main_Page。

②OpenStack－开源中国社区：

· https://www. oschina. net/question/tag/openstack。

· 开源中国社区:http://www. oschina. net。

③中外文的相关资源。

在搜索引擎上查找 OpenStack 核心模块对应的网站或 Blog。

（2）OpenStack 的项目案例。

以下基于 OpenStack 的 2019 年全国云计算技术与应用技能竞赛考核知识结构和相关内容 OpenStack 项目的实施环节内容要求。

案例名称：2019 年全国云计算技术与应用技能竞赛。

项目要求：基于 OpenStack 的云计算服务平台的搭建。

项目的具体考核与考核知识及技能点如表1-1所示。表中的有些内容并不属于课程的教学内容，可帮助读者了解云计算服务平台上的服务与应用项目内容。

表1-1 基于OpenStack的云计算技术与应用技能竞赛考核要求

考核环节	考核知识点和技能点
IaaS云计算基础架构平台	按照系统网络架构要求，检查网络设备和服务器设备连线、配置是否正确
	CentOS Linux操作系统检查，配置主机名，配置YUM安装源。通过系统的配置文件检查正确性
	基本服务SELinux、NTP、FTP、MySQL、MongoDB、rabbitMQ的安装、配置和使用。通过配置文件或验证命令查看正确性
	使用部署脚本，快速部署IaaS平台Keystone安全统一框架服务。通过配置文件或验证命令查看正确性
	使用部署脚本，快速部署IaaS平台镜像Glance服务。通过配置文件或验证命令查看正确性
	使用部署脚本，快速部署IaaS平台计算Nova服务。通过配置文件或验证命令查看正确性
	使用部署脚本，快速部署IaaS平台网络Neutron服务。通过配置文件或验证命令查看正确性
	使用部署脚本，快速部署IaaS平台控制面板Horizon，管理云平台虚拟交换机。通过配置文件或验证命令查看正确性
	使用部署脚本，快速部署IaaS平台块存储Cinder。通过配置文件或验证命令查看正确性
	使用部署脚本，快速部署IaaS平台对象存储Swift服务。通过配置文件或验证命令查看正确性
	使用部署脚本，快速部署IaaS平台模板Heat服务。通过配置文件或验证命令查看正确性
	使用部署脚本，快速部署IaaS平台监控Ceilometer和报警Alarm服务。通过配置文件或验证命令查看正确性
	规划和构建SDN OpenDaylight云网络，建立统一的云计算平台网络管理和服务架构
	使用部署脚本，快速部署IaaS平台云数据库Trove服务。通过配置文件或验证命令查看正确性
	通过云平台提供的对外restful接口对云平台的服务进行增、删、查、改的操作。完成后通过管理命令行验证正确性
	使用部署文档，部署日志分析服务，通过配置文件或验证命令查看正确性
	使用部署文档，部署入侵检测服务，通过配置文件或验证命令查看正确性
	使用部署脚本，快速部署防火墙与负载均衡服务，通过配置文件或验证命令查看正确性
	使用部署文档，部署数据库高可用案例，通过配置文件或验证命令查看正确性
	使用部署文档，部署监控nagios监控服务，通过配置文件或验证命令查看正确性
	通过部署文档，部署https访问，通过配置文件或验证命令查看正确性
PaaS云计算开发服务平台	修改系统配置部署Docker Engine，完成后通过上传镜像Image进行测试和验证，通过配置文件或验证服务命令查看正确性
	搭建本地镜像仓库Image Repositories，部署和配置Docker Registry服务，搭建完成后通过配置文件或验证命令查看正确性
	通过对镜像和容器查询、使用和管理，并通过查询网络、存储等信息验证容器的正确性
	部署和配置Docker Compose容器编排服务，搭建完成后，使用Compose编排构建应用进行验证
	部署和配置Rancher构建容器服务（CaaS），搭建完成后，通过创建容器或应用验证正确性

续表

考核环节	考核知识点和技能点
云计算平台运维管理	管理 IaaS 平台 MySQL 数据库、rabbitMQ 消息服务、MongoDB 数据库服务和运行日志。通过排错和后台监控，提交系统运作状态
	管理 IaaS 底层服务包括 LVM、OVS、网桥、KVM 等服务。通过日志排错和后台监控，提交系统运作状态
	管理 IaaS 平台 Keytone 认证，使用命令和管理员界面，为企业创建租户和用户。通过查询数据库、日志排错和后台监控进行验证，查看正确性
	使用 Glance 服务，制作 Window、Ubuntu 镜像，使用镜像部署云主机，通过命令的方式查看云主机的状态信息
	管理 IaaS 平台网络 Neutron 服务，使用云平台网络服务，配置不同的网络模式：Flat、GRE、VLAN，完成不同网络模型的配置。配置 L3 . LB、DVR 的网络扩展支持。通过管理命令、日志排错和后台监控验证正确性
	管理 IaaS 平台 Cinder 块存储服务，为云主机挂载虚拟硬盘，对云平台的数据进行同步灾备，创建加密块设备，保证数据安全。通过使用、管理命令验证正确性
	管理 IaaS 平台 Swift 对象存储服务，使用和管理账户、容器和对象，完成一个网盘存储场景的构建。完成后提交配置参数，使用、管理命令验证正确性
	基于 Ceph 构建 IaaS 平台统一云存储，分别支撑 Glance、Cinder、Swift 云存储后端，完成后提交配置参数，通过管理命令、日志排错和后台监控验证正确性
	管理 IaaS 数据库 Trove 服务，进行支撑 MySQL、Cassandra、MongoDB 配置和使用。完成后提交配置参数，通过使用、管理命令验证正确性
	管理 IaaS 监控 Ceilometer 服务，通过管理命令或管理界面，查看云平台各服务、实例、存储和网络的运行状态
	管理 IaaS 模板 Heat 服务，使用模板服务，按照模板标准，定义生产系统的云主机模板，并通过管理命令或管理界面上传模板。完成后使用模板创建云主机并提交云主机状态
	对 IaaS 平台进行基本服务的云主机、云存储、云网络的系统错误的排查。完成任务后，提交排查的问题和正确运行结果
	系统上云综合案例，设计和构建 Web 系统上云，申请云主机，配置云数据库，配置云存储，配置负载均衡
	基于入侵检测服务，配置日志保存路径，查看请求 log 与响应 log
	基于日志分析服务，修改日志读/写权限，创建索引，添加字段分析，分析云平台各服务组件日志是否有异常
	基于 nagios 监控服务，监控云平台各项指标，如 CPU 使用率、内存使用率等
	基于防火墙和负载均衡服务，配置防火墙规则和负载均衡协议，并查询结果
	基于数据库高可用服务，查询数据库同步状态，导入数据，验证数据同步结果
	容器基础技术 CGroup 和 NameSpace 的使用和运维，通过管理命令测试、验证正确性
	根据需求定义 Dockerfile 镜像模板，上传并运行测试，完成后通过测试、验证命令查看正确性
	对 Docker 的存储、数据卷、网络进行配置和管理，使用 Docker 命令进行镜像、容器的操作和运维。通过使用、管理命令、日志排错和后台监控验证正确性

续表

考核环节	考核知识点和技能点
云计算平台运维管理	使用PaaS平台,构建软件服务:包括Web服务器(Nginx)、数据库(MongoDb、Mysql)、代码管理系统(Gogs)、搜索引擎(Elasticsearch 2.x)、持续集成(Jenkins)、监控系统(Grafana、Prometheus)等。通过配置文件或验证命令查看正确性
	对PaaS平台进行基本服务的镜像、容器、存储、网络的系统错误的排查。完成任务后,提交排查的问题和正确运行结果
	实现对PaaS平台应用商店增加新应用功能,启动应用并验证
大数据平台	Ambari分布式平台管理工具的安装、配置和使用,主要包括数据库、ambari - server和ambari - agent运维管理,安装完成后对大数据平台的系列服务进行统一部署、管理和监控
	Hadoop HDFS和Map - Reduce的配置和使用,通过运行案例验证Map - Reduce,对HDFS文件系统进行运维操作
	数据仓库Hive配置和应用,使用Hive进行数据仓库的增、删、查、改和管理的运维操作
	分布式列数据库HBase配置和应用,使用HBase进行分布式列数据库的增、删、查、改和管理的运维操作
	数据挖掘工具Mahout配置和应用,使用Mahout进行数据挖掘分析
	Pig大数据处理工具的配置和应用,部署成功后使用Pig进行数据处理
	Sqoop数据库传输工具的配置和应用,部署成功使用Sqoop进行数据库间的数据传输
	Flume日志采集工具,部署成功后使用Flume进行日志收集,分析
	Spark内存运算分布式框架的配置和应用,使用Spark进行案例分析
	实现Ambari平台增加新服务功能,在页面添加服务并验证服务安装成功
SaaS云应用开发	导入大数据框架项目,并正确配置
	配置基础环境,包括MySQL、MongoDB、HBase等,并测试连接成功
	基于给定的大数据源、大数据服务进行数据采集开发
	基于给定的大数据源、数据处理模型,进行数据处理应用开发
	基于给定的大数据源、数据处理模型,进行数据分析应用开发
	根据给定的要求,进行订单的界面和功能开发
	根据给定的要求,进行订单详情的界面和功能开发
	根据给定的要求,进行商品详情界面和功能开发
	根据给定的要求,进行用户管理界面和功能开发
工程文档及职业素养	工程文档编写,编写平台设计文档、配置文件、架构图和测试报告
	工程文档编写,编写Shell运维脚本、Linux常见运维命令与服务、云平台基础错误解析、功能模块的系统流程图、程序UML图等
	比赛现场符合企业"5S"(即整理、整顿、清扫、清洁和素养)原则
	团队分工明确合理、操作规范、文明竞赛

任务二　云计算平台的系统架构设计

任务要求

小张基本掌握云计算平台搭建的基础知识，接下来需要对公司的应用需求进行调研，在此基础上要进行公司云计算平台的系统环境设计和系统搭建安装工作。

核心概念

云计算平台的架构：依据 OpenStack 架构指南，遵循 IaaS 模式，搭建基于简单的需求为用户寻求最合适的云计算平台架构，包括网络拓扑结构和系统架构。

知识准备

视频
云平台的需求及架构设计

经过调研分析，公司的情况及对云平台应用的需求如下：

（1）公司的基本组织结构。公司内部有 10 名员工，其中 2 名在项目研发部（研发环境），3 名在业务部（办公环境），5 人在 IT 工程部（运维环境）。根据企业人员部门分配，现构建 3 个租户、10 个用户，管理人员拥有管理员权限，其余人员拥有普通用户权限，如图 1-4 所示。

图 1-4　公司的基本结构

（2）应用需求情况。公司的应用需求如图 1-5 所示。

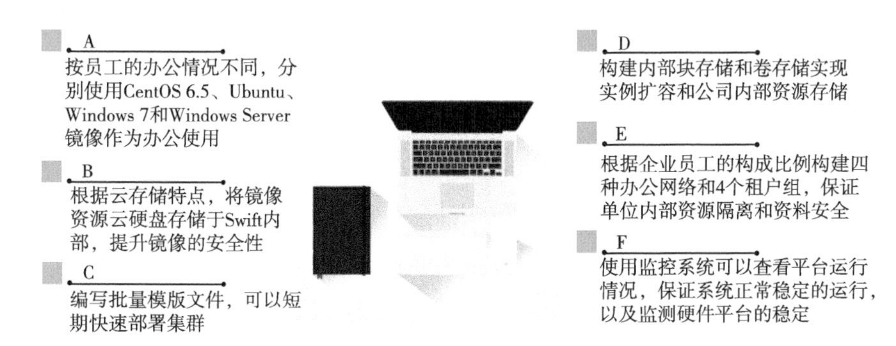

A
按员工的办公情况不同，分别使用CentOS 6.5、Ubuntu、Windows 7和Windows Server镜像作为办公使用
B
根据云存储特点，将镜像资源云硬盘存储于Swift内部，提升镜像的安全性
C
编写批量模版文件，可以短期快速部署集群
D
构建内部块存储和卷存储实现实例扩容和公司内部资源存储
E
根据企业员工的构成比例构建四种办公网络和4个租户组，保证单位内部资源隔离和资料安全
F
使用监控系统可以查看平台运行情况，保证系统正常稳定的运行，以及监测硬件平台的稳定

图 1-5　公司的应用需求

（3）服务需求。公司对云平台的服务需求，如图1-6所示。

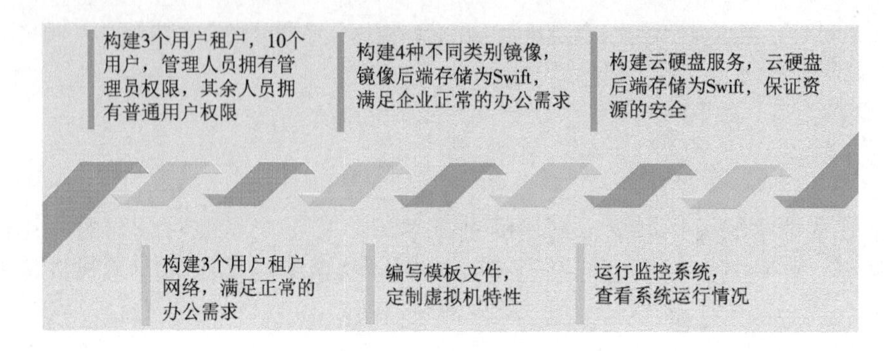

图1-6　公司对云平台的服务需求

1. 云平台系统架构设计

按照既定的项目目标，接下来围绕这个目标开始一步一步构建云计算平台，满足日常的企业办公、生产和研发，基于以上要求，依据OpenStack架构指南，构建一个通用性云平台，遵循IaaS模式，基于简单的需求为用户寻求最合适的平台。

基于简单的需求为用户寻求最合适的平台。通用性为最基本、最简单的平台，适合概念验证、小型实验，也可以基于通用性平台随意扩展计算资源和存储资源。以下就是整套平台环境的网络拓扑结构的说明。

1）拓扑说明

在云平台的网络拓扑结构图中，采用两种节点服务器构建云计算平台，其中一种为控制节点服务器，另一种为计算节点（即实例节点）服务器。按照网络分离和功能化要求，也依次构建了3种网络，分别为实例通信网络、内部管理网络和实例私有网络。同时考虑到服务器只有两个网口的实际情况，采取结合Open vSwitch虚拟交换机功能虚拟生成3个网口，对应为br-ex、br-mgmt、br-prv，分别作为实例通信网口、内部管理网络网卡、实例私有网络网卡。将服务器流量转化为两个交换机将数据流量分流：管理交换机为管理流量，数据交换机为实例私有和实例通信网络流量，如图1-7所示。

2）系统架构设计

根据系统拓扑介绍，每个节点模块的安装服务也是根据此来进行确定。根据前期系统的部署示意将平台服务做了拆分，节点部署服务示意图如图1-8所示。

控制节点有消息服务（RabbitMQ）、数据库服务（MySQL）、认证服务（KeyStone）、镜像服务（Glance）、计算控制服务（Nova）、网络控制服务（Neutron）、控制台服务（Horizon）、块存储控制服务（Cinder）、对象存储控制服务（Swift）、编配服务（Heat）和监控服务（Ceilometer）组件，计算节点有计算控制服务（Nova）、网络控制服务（Neutron）和监控服务（Ceilometer）。

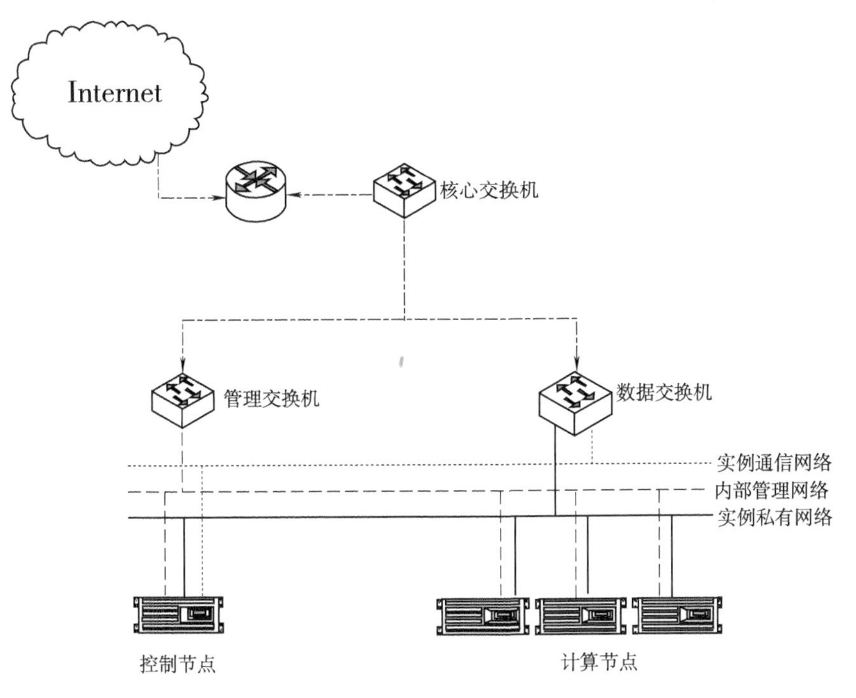

图 1-7　云平台的网络拓扑结构

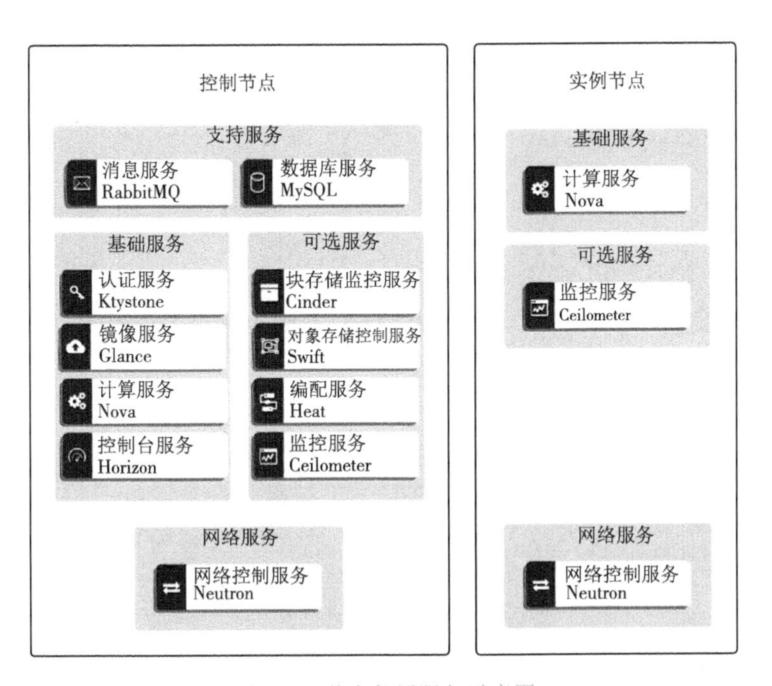

图 1-8　节点部署服务示意图

2. 公司私有云平台系统架构设计

根据节点部署服务示意图，公司私有云平台系统架构可以设计成控制节点与计算节点，且需要控制节点与计算节点安装以下的组件。

1）控制节点安装服务

消息服务（RabbitMQ）、数据库服务（MySQL）、认证服务（Keystone）、镜像服务（Glance）、计算控制服务（Nova）、网络控制服务（Neutron）、控制台服务（Horizon）、块存储控制服务（Cinder）、对象存储控制服务（Swift）、编配服务（Heat）和监控服务（Ceilometer），完成云平台控制端的安装工作。

2）计算节点安装服务

计算控制服务（Nova）、网络控制服务（Neutron）和监控服务（Ceilometer），完成计算节点的安装。

任务三　云平台系统基础安装

任务要求

小张在上述系统架构的设计基础上开展下一步的工作，准备好 OpenStack 搭建云计算平台项目所需的软件资源包、按云平台的网络拓扑结构图进行设备准备与网络连接，完成搭建云计算基础服务。

核心概念

云平台系统安装基础工作：包括准备 OpenStack 搭建云平台项目所需的软件资源包、确定各节点的名称、配置各节点的 IP 网络地址、按要求安装各节点的操作系统、配置系统环境变量、在控制节点和计算节点分别运行脚本、完成各节点的配置安装及验证安装基础工作。

知识准备

1. 节点主机名及 IP 地址规划

根据拓扑结构图，本次部署的节点主机名及 IP 地址分配如表 1－2 所示，同时，各节点的主机名称也规划在表中。

表 1－2　各节点主机名和 IP 地址规划列表

节点 主机	主机名	IP 规划		
		实例通信	内部管理	内部私有
控制节点	controller	192.168.200.10	192.168.100.10	192.168.400－600.0/24
计算节点	compute	192.168.200.20	192.168.100.20	192.168.400－600.0/24

2. 各节点的安装系统要求

各节点对主机、系统及节点的要求如图 1-9 所示。其中，节点主要指部署云平台的物理节点服务器。控制节点：存放系统数据库、中间件服务，实际为云平台系统的大脑和控制中心。计算节点：存放虚拟机的服务器，支持处理器虚拟化功能，运行虚拟机管理程序（QEMU 或 KVM）管理虚拟机主机，同时为外部用户提供存储服务和内部实例提供块存储服务。

图 1-9　各节点对主机、系统及节点的要求

视频 Linux相关的操作系统知识

3. 与 Linux 相关的操作知识

（1）OpenStack 云计算平台的搭建过程中需要重点知道的一些基础操作知识。

①Linux 的版本可分为内核版本与发行版本。内核版本是 Linux 任何版本的核心，内核版本号由 3 个数字组成：X. Y. Z。发行版本是将 Linux 内核与应用软件打包发行的版本，主流的 Linux 发行版本有：Red Hat Enterprise Linux 、CentOS、Ubuntu、Debian GNU/Linux、openSUSE、红旗 Linux 等。

这里 OpenStack 云计算平台的搭建使用 CentOS 7. 2_x64bit 版本。

②云服务器的安装是要求服务器中，包含处理器模块、存储模块、网络模块、电源、风扇等设备。云服务器关注的是高性能吞吐量计算能力，关注的是在一段时间内的工作总和。因此，云服务器在架构上和传统的服务器有着很大的区别，具有庞大的数据输入量或海量的工作集。

这里 OpenStack 云计算平台需要购买带有双网卡的高性能服务器或者也可以用高配置计算机，加两块网卡。

③云服务器的硬盘分区要求如下：对控制节点的硬盘分区没有特殊的要求，但是，对计算节点一般可预留两个空分区，如图 1-10 所示。

Please Select A Device

Device	Size (MB)	Mount Point/ RAID/Volume	Type	Format
▽ LVM Volume Groups				
▽ vg_controller	21096			
LogVol01	19048	/	ext4	✓
LogVol00	2048		swap	✓
▽ Hard Drives				
▽ sda				
sda1	100	/boot	ext4	✓
sda2	20000		ext4	✓
sda3	10000		ext4	✓
▽ sda4	21099		Extended	
sda5	21098	vg_controller	physical volume (LVM)	✓

Create　Edit　Delete　Reset

Back　Next

图 1-10　计算节点的硬盘分区

④云服务器安装 Linux 系统时需要注意以下几方面：

· 在进行安装时请选择英文界面，填写主机名称（Hostname）。

· 单击 Configure Network 作 IP 地址的设置。

· 不选 System clock uses UTC。

· 选择 Create Custom Layout 进行系统分区，在选择安装系统时可选最小化（Mini）安装。

（2）系统配置文件。常用的系统配置文件及其功能如表 1－3 所示。

表 1－3　常用的系统配置文件及其功能

序号	配置文件	所在子目录	功能
1	hosts	/etc	主机名与 IP 地址的映射关系
2	Network	/etc/sysconfig/	主机名称
3	ifcfg－eth0	/etc/sysconfig/network－scripts/	网卡 0 的 IP 地址
4	Config	/etc/selinux/	selinux 的配置
5	Iptales	/etc/sysconfig/	配置防火墙规则

1. 准备实施任务的环境（见表 1－4）

表 1－4　实施任务的准备环境

实施任务所需软件资源	虚拟机镜像资源信息
虚拟机软件(vm12)	CentOS－7－x86_64－DVD－1511.iso
securcrt 远程连接软件	Xiandian－IaaS－2.2.iso

2. 分解实施任务的过程（见表 1－5）

表 1－5　实施任务的简明过程

序号	步骤	详细操作及说明
1	准备 OpenStack 搭建云计算平台项目所需的软件资源包	
2	确定各节点的名称，控制节点名称为 controller，计算节点名称为 compute	
3	确定控制节点和计算节点配置各节点的 IP 网络地址	
4	按要求安装各节点的操作系统	

续表

序号	步骤	详细操作及说明
5	控制节点的基础部署	①配置修改控制节点的主机名； ②在全部节点的/etc/hosts 文件中添加域名解析； ③配置环境、配置防火墙规则； ④yum 安装； ⑤配置 IP； ⑥控制节点虚拟机安装 iaas – xiandian 安装包； ⑦在控制节点虚拟机上安装 iaas – pre – host； ⑧部署脚本安装平台
6	计算节点的基础部署	①配置修改计算节点的主机名； ②在全部节点的/etc/hosts 文件中添加域名解析； ③配置环境、配置防火墙规则； ④yum 安装； ⑤配置 IP； ⑥计算节点虚拟机安装 iaas – xiandian 安装包； ⑦在计算节点虚拟机上安装 iaas – pre – host； ⑧部署脚本安装平台
7	验证安装基础工作完成	

3. 实现云平台基础部署工作任务

（1）准备 OpenStack 搭建云计算平台项目所需的软件资源包。

它们分别是 centos – 7 – x86_64 – DVD – 1511. iso、Xiandian – IaaS – 2. 2. iso、虚拟机软件（vm12）及 securcrt 远程连接软件。其中，Xiandian – IaaS – 2. 2 镜像包是来自南京第五十五所技术开发有限公司的先电品牌 CLOUD – TR200 型号的产品之一，它是基于以 Apache 开放许可证授权开源云计算项目 OpenStack 开发的，可管理主流的 Hypervisor（VMware vSphere、微软 Hyper – V、Citrix XenServer 、KVM、Xen、VirtualBSD）。全部软件资源包如图 1– 11 所示。

CentOS-7-x86_64-DVD-1511	2018/5/18 7:54	好压 ISO 压缩文件	4,228,096...
securcrt	2018/10/14 12:36	好压 RAR 压缩文件	23,983 KB
VMware-workstation-full-12.5.7-581...	2019/1/15 7:08	应用程序	414,641 KB
XianDian-IaaS-v2.1	2018/4/28 17:10	好压 ISO 压缩文件	2,779,556...
XianDian-IaaS-v2.2	2017/11/6 11:10	好压 ISO 压缩文件	2,784,670...

图 1– 11　全部软件资源包

（2）确定各节点的名称，控制节点名称为 controller，计算节点名称为 compute，如表 1 – 6 所示。

表 1－6　节点名称

节点名称	中文名
controller	控制节点
compute	计算节点

（3）确定控制节点和计算节点配置各节点的 IP 地址，如表 1－7 所示。

表 1－7　各节点的 IP 地址

节点名称	网卡	IP 地址
controller	第一张网卡	192. 168. 100. 10
	第二张网卡	192. 168. 200. 10
compute	第一张网卡	192. 168. 100. 20
	第二张网卡	192. 168. 200. 20

（4）新建虚拟机。

新建两个虚拟机，分别名为 controller 节点和 compute 节点，它们的配置如表 1－8所示。新建控制节点虚拟机的操作如图 1- 12 ~ 图 1- 16 所示。计算节点的虚拟机类似。

表 1－8　虚拟机的配置

虚拟机名称	CPU	内存	存储	网卡	备注
controller	处理器双核	4 GB	500 GB	两张网卡，分别是 VNet1/VNet8	开启 CPU 虚拟机
compute	处理器四核	4 GB	500 GB	两张网卡，分别是 VNet1/VNet8	开启 CPU 虚拟机

图 1- 12　选择客户机操作系统

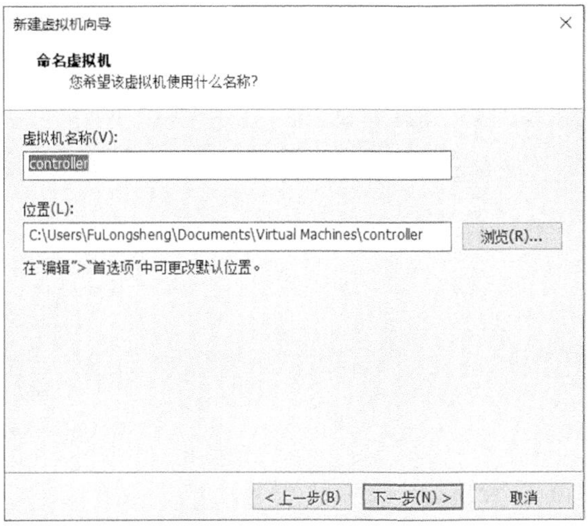

图 1-13　命名虚拟机

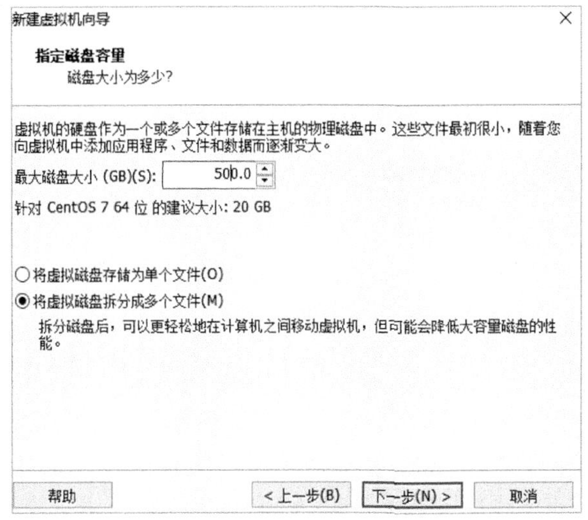

图 1-14　指定磁盘容量

图 1-15　指定 CPU 和内存大小，并选择“虚拟化 Intel VT－x/EP T 或 AMD－V/RVI（V）”选框

图 1-16　设置两张网络适配器

(5) 按要求安装各节点的操作系统。

新建控制节点和计算节点两台虚拟机后安装 CentOS7.2，图 1-17 所示是各节点的系统安装选择安装语言为中文；图 1-18 所示为各节点的系统安装选择安装信息，安装位置选择分区为自定义；图 1-19 所示是各节点的系统安装手动分区的信息；图 1-20所示为各节点的系统安装的进度。

图 1-17　各节点的系统安装选择安装语言

安装信息摘要
CENTOS 7 安装
cn
Help!
CentOS
SECURITY
SECURITY POLICY
No profile selected
软件
安装源(I)
本地介质
软件选择(S)
最小安装
系统
安装位置(D)
已选择自定义分区
KDUMP
已禁用 Kdump
网络和主机名(N)
未连接
退出(Q)
开始安装(B)
在点击"开始安装"按钮前我们并不会操作您的磁盘

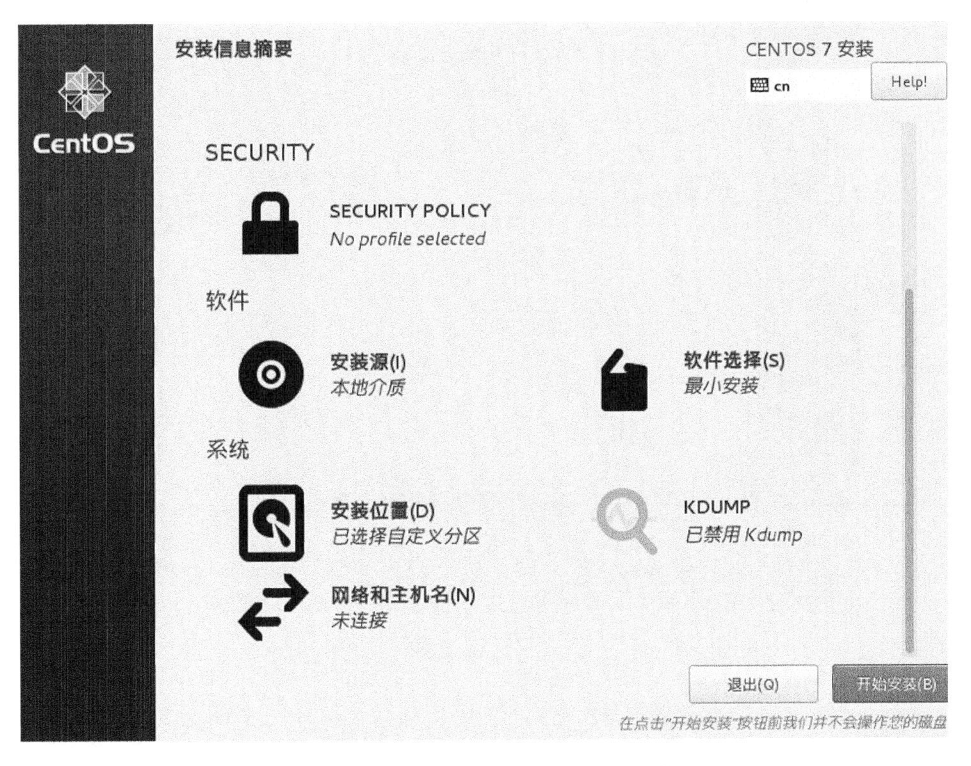

图 1-18　各节点的系统安装选择安装信息

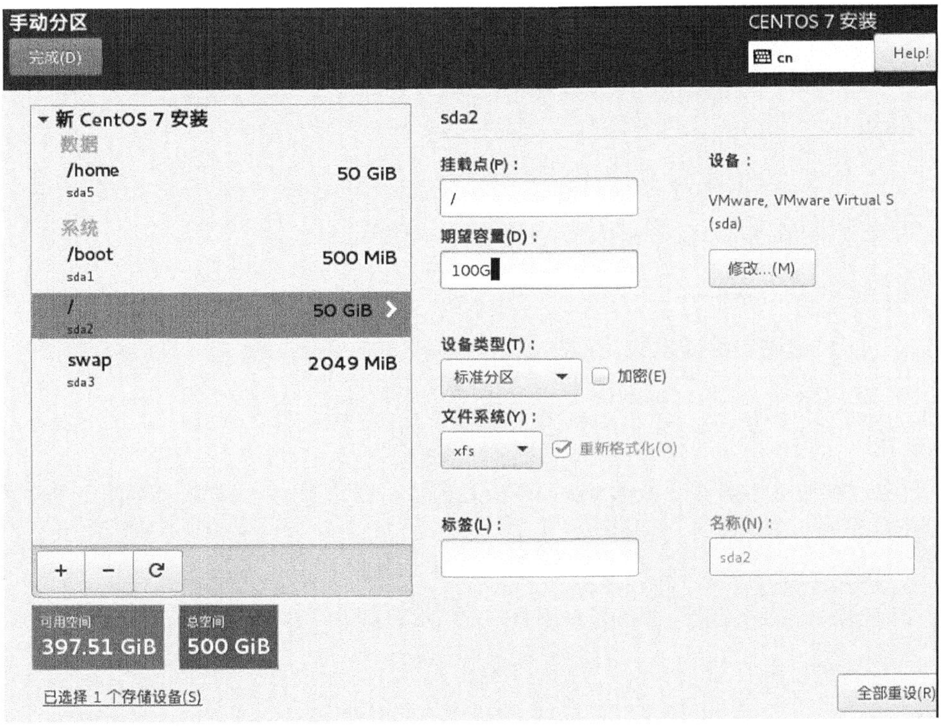

手动分区
完成(D)
CENTOS 7 安装
cn
Help!
新 CentOS 7 安装
数据
/home
50 GiB
sda5
系统
/boot
500 MiB
sda1
/
50 GiB
sda2
swap
2049 MiB
sda3
sda2
挂载点(P)：
/
期望容量(D)：
100G
设备：
VMware, VMware Virtual S
(sda)
修改...(M)
设备类型(T)：
标准分区
加密(E)
文件系统(Y)：
xfs
重新格式化(O)
标签(L)：
名称(N)：
sda2
可用空间
397.51 GiB
总空间
500 GiB
已选择 1 个存储设备(S)
全部重设(R)

图 1-19　各节点的系统安装手动分区信息

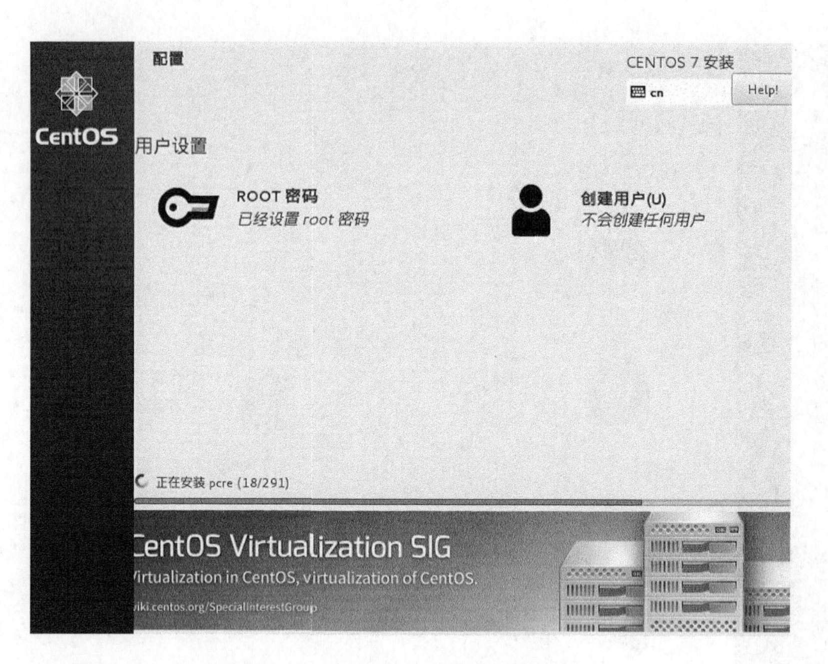

图 1-20　各节点的系统安装的进度

视频 控制节点的安装

（6）控制节点的基础部署。

①配置修改控制节点的主机名。

```
[root@ localhost ~] #  hostnamectl set -
hostnamecontroller
```

修改主机名执行成功后没有返回结果。

通过 bash 刷新，主机名会变为 controller。

```
[root@ localhost ~] #  bash
```

②在控制节点的/etc/hosts 文件中添加域名解析。

```
[root@ controller ~] #  vi /etc/hosts
```

按【I】键进入编译模式，添加以下内容：

```
192.168.100.10 controller
192.168.100.20 compute
```

注：在最后添加自己虚拟机的 IP 和主机名，注意 IP 和主机名之间有一个空格。按【Esc】键并输入：wq 命令保存退出。

③配置环境配置防火墙规则。

清除所有 chains 链（INPUT/OUTPUT/FORWARD）中所有的 rule 规则。

```
[root@ controller ~] #iptables -F
```

清空所有 chains 链（INPUT/OUTPUT/FORWARD）中包及字节计数器。

```
[root@ controller ~] #iptables -X
```

清除用户自定义 chains 链（INPUT/OUTPUT/FORWARD）中的 rule 规则。

```
[root@ xiandian ~] # iptables -Z
```

执行清除命令没有返回结果，通过/usr/sbin/iptables - save 保存。

```
[root@ controller ~] #  /usr/sbin/iptables - save
```

执行结果如图 1-21 所示。

```
[root@xiandian ~]# /usr/sbin/iptables-save
# Generated by iptables-save v1.4.21 on Thu May 30 05:01:11 2019
*filter
:INPUT ACCEPT [67:4892]
:FORWARD ACCEPT [0:0]
:OUTPUT ACCEPT [45:3844]
COMMIT
# Completed on Thu May 30 05:01:11 2019
```

图 1-21　执行结果

④ yum 安装。

a. 在控制节点挂载镜像制作源路径。

挂载 CentOS - 7 - x86_64 - DVD - 1511. iso：

```
[root@ controller ~] #  mount -o loop /dev/cdrom  /mnt/
[root@ controller ~] #  mkdir /opt/centos
[root@ controller ~] #  cp -rf /mnt/* /opt/centos/
[root@ controller ~] #  umount  /mnt/
```

挂载 Xiandian - IaaS - v2. 2. iso：

```
[root@ controller ~] #  mount -o loop /dev/cdrom/mnt/
[root@ controller ~] #  cp -rf /mnt/* /opt/
[root@ controller ~] #  umount  /mnt/
```

挂载情况如图 1-22 所示。

```
[root@xiandian ~]# mount -o loop CentOS-7-x86_64-DVD-1511.iso /mnt/
mount: /dev/loop0 is write-protected, mounting read-only
[root@xiandian ~]# mkdir /opt/centos
[root@xiandian ~]# cp -rf /mnt/* /opt/centos/
[root@xiandian ~]#
[root@xiandian ~]# umount  /mnt/
[root@xiandian ~]#
[root@xiandian ~]# mount -o loop XianDian-IaaS-v2.2.iso  /mnt/
mount: /dev/loop0 is write-protected, mounting read-only
[root@xiandian ~]# cp -rf /mnt/* /opt/
[root@xiandian ~]#
[root@xiandian ~]# umount  /mnt/
```

图 1-22　挂载情况

b. 配置 yum 路径。

将网络 yum 源路径移除 yum 目录。

```
[root@ controller ~] #  mv /etc/yum. repos. d/* /opt
```

执行没有返回结果。

c. 在控制节点创建 repo 文件。

在/etc/yum. repos. d 创建 centos. repo 源文件。

```
[root@ controller ~] # vi /etc/yum. repos. d/local. repo
```

按【I】键进入编译模式，添加以下内容到 local. repo 文件里：

```
[centos]
name = centos
baseurl = file: ///opt /centos
gpgcheck = 0
enabled = 1
[iaas]
name = iaas
baseurl = file: ///opt /iaas - repo
gpgcheck = 0
enabled = 1
```

按【Esc】键并输入：wq 命令保存退出。

输入测试命令：

```
[root@ controller ~] # yum repolist
```

⑤配置 IP。

根据服务器自身的 IP 地址配置网卡信息。

```
[root@ controller ~] #  vi /etc /sysconfig /network - scripts /
ifcfg - eno16777736
```

按【I】键进入编译模式，将文件修改为以下形式：

```
TYPE = Ethernet
BOOTPROTO = static
NM_ CONTROLLED = yes
DEVICE = eno1677773
ONBOOT = yes
IPADDR = 192. 168. 100. 10
PREFIX = 24
GATEWAY = 192. 168. 100. 1
```

按【Esc】键并输入：wq 命令保存退出。

注：将获取 IP 方式改为静态。

在 IPADDR 输入自己虚拟机的 IP 地址。

网关也是结合服务器所在网段填写。

```
[root@ controller ~] #  vi /etc /sysconfig /network - scripts /
ifcfg - eno33554960
```

按【I】键进入编译模式，将文件修改为以下形式：

```
TYPE = Ethernet
BOOTPROTO = static
NM_ CONTROLLED = yes
DEVICE = eno33554960
ONBOOT = yes
IPADDR =192 . 168 . 200 . 10
PREFIX =24
NETGATE =192 . 168 . 200 . 1
```

按【Esc】键并输入：wq 命令保存退出。

注：将获取 IP 方式改为静态。

在 IPADDR 输入自己虚拟机的 IP 地址。

外网段不能和虚拟机 IP 在同一个网段。

重启网络：

```
[root@ controller ~] # systemctl restart network
```

⑥控制节点虚拟机安装 iaas - xiandian 安装包。

```
[root@ controller ~] # yum install  -y iaas - xiandian
```

下载情况如图 1-23 所示。

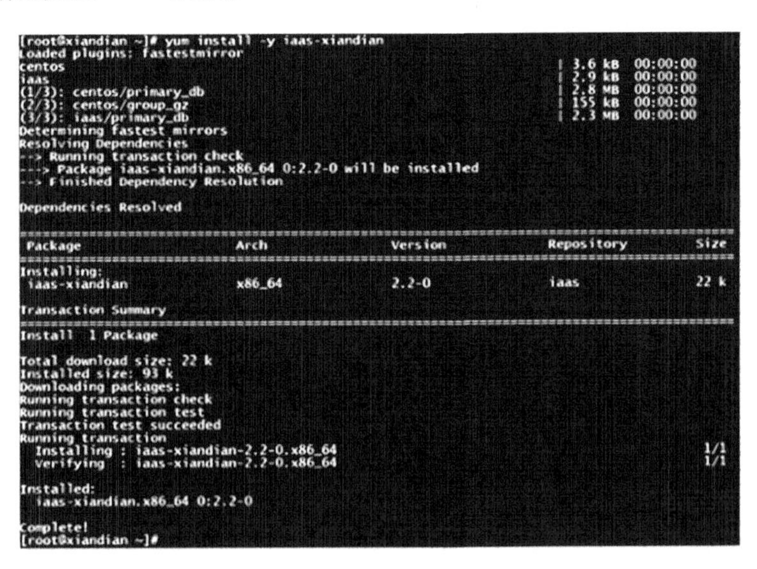

图 1-23　下载情况

编辑文件/etc/xiandian/openrc. sh(配置环境变量)。

```
[root@ controller ~] # vi /etc/xiandian/openrc. sh
```

按【I】键进入编译模式，按以下参数修改配置文件。

说明： 原配置文件中是有“#”号的，填写好配置后删除注释“#”。

```
HOST_IP=192.168.100.10
HOST_NAME=controller
HOST_IP_NODE=192.168.100.20
HOST_NAME_NODE=compute
RABBIT_USER=openstack
RABBIT_PASS=000000
DB_PASS=000000
DOMAIN_NAME=demo(自定义)
ADMIN_PASS=000000
DEMO_PASS=000000
KEYSTONE_DBPASS=000000
GLANCE_DBPASS=000000
GLANCE_PASS=000000
NOVA_DBPASS=000000
NOVA_PASS=000000
NEUTRON_DBPASS=000000
NEUTRON_PASS=000000
METADATA_SECRET=000000
INTERFACE_NAME=eno33554960
CINDER_DBPASS=000000
CINDER_PASS=000000
BLOCK_DISK=vda3
SWIFT_PASS=000000
OBJECT_DISK=vda4
STORAGE_LOCAL_NET_IP=127.0.0.1
HEAT_DBPASS=000000
HEAT_PASS=000000
CEILOMETER_DBPASS=000000
CEILOMETER_PASS=000000
AODH_DBPASS=000000
AODH_PASS=000000
```

按【Esc】键并输入：wq 命令保存退出。

⑦在控制节点虚拟机上安装 iaas-pre-host，安装好后重启设备，重启会卡住需要单击关机并退出按钮，退出虚拟机，然后再次进入实验。

```
[root@controller ~]#iaas-pre-host.sh
[root@controller ~]#reboot
```

注：如果没有反应，请重新刷新平台页面。

⑧部署脚本安装平台。

在控制节点虚拟机执行脚本 iaas-install-mysql.sh 进行数据库及消息列表服务安装。

```
[root@ controller ~] # iaas - install - mysql.sh
```

在控制节点虚拟机执行脚本 iaas - install - keystone. sh 进行 keystone 认证服务安装。

```
[root@ controller ~] # iaas - install - keystone.sh
```

在控制节点虚拟机执行脚本 iaas - install - glance. sh 进行 glance 镜像服务安装。

```
[root@ controller ~] # iaas - install - glance.sh
```

在控制节点虚拟机执行脚本 iaas - install - nova - controller. sh 进行 nova 计算服务安装。

```
[root@ controller ~] # iaas - install - nova - controller.sh
```

在控制节点虚拟机执行脚本 iaas - install - neutron - controller. sh 进行 neutron 网络服务安装。

```
[root@ controller ~] # iaas - install - neutron - controller.sh
```

在控制节点虚拟机执行脚本 iaas - install - neutron - controller - gre. sh 进行 gre 网络安装配置。

```
[root @ controller ~] # iaas - install - neutron - controller -
gre.sh
```

⑨在控制节点虚拟机执行脚本 iaas - install - dashboard. sh 进行 Dashboard 服务安装。

```
[root@ controller ~] # iaas - install - dashboard.sh
```

(7)计算节点的基础部署

①配置计算节点的主机名。

```
[root@ localhost ~] # hostnamectl set -
hostnamecompute
```

修改主机名执行成功后没有返回结果。

通过 bash 刷新,主机名会变为 compute。

```
[root@ compute ~] # bash
```

②在计算节点的/etc/hosts 文件中添加域名解析。

```
[root@ compute ~] # vi /etc /hosts
```

按【I】键进入编译模式,添加以下内容:

```
192.168.100.10 controller
192.168.100.20 compute
```

③ 配置环境配置防火墙规则。

清除所有 chains 链(INPUT/OUTPUT/FORWARD)中所有的 rule 规则。

```
[root@ compute ~]  # iptables -F
```

清空所有 chains 链(INPUT/OUTPUT/FORWARD)中包及字节计数器。

```
[root@ comput e ~]  # iptables -X
```

清除用户自定义 chains 链(INPUT/OUTPUT/FORWARD)中的 rule 规则。

```
[root@ compute ~]  # iptables -Z
```

执行清除命令没有返回结果,通过/usr/sbin/iptables - save 保存。

```
[root@ compute ~]  # /usr/sbin/iptables-save
```

执行结果如图 1-24 所示。

```
[root@xiandian ~]# /usr/sbin/iptables-save
# Generated by iptables-save v1.4.21 on Thu May 30 05:01:11 2019
*filter
:INPUT ACCEPT [67:4892]
:FORWARD ACCEPT [0:0]
:OUTPUT ACCEPT [45:3844]
COMMIT
# Completed on Thu May 30 05:01:11 2019
```

图 1-24　执行结果

④yum 安装。

a. 在计算节点挂载镜像制作源路径。

挂载 CentOS-7-x86_64-DVD-1511.iso:

```
[root@ compute ~]  # mount -o loop /dev/cdrom /mnt/
[root@ compute ~]  # mkdir /opt/centos
[root@ compute ~]  # cp -rf /mnt/* /opt/centos/
[root@ compute ~]  # umount  /mnt/
```

挂载 Xiandian-IaaS-v2.2.iso:

```
[root@ compute ~]  # mount -o loop /dev/cdrom /mnt/
[root@ compute ~]  # cp -rf /mnt/* /opt/
[root@ compute ~]  # umount  /mnt/
```

挂载情况如图 1-25 所示。

```
[root@xiandian ~]# mount -o loop CentOS-7-x86_64-DVD-1511.iso /mnt/
mount: /dev/loop0 is write-protected, mounting read-only
[root@xiandian ~]# mkdir /opt/centos
[root@xiandian ~]# cp -rf /mnt/* /opt/centos/
[root@xiandian ~]#
[root@xiandian ~]# umount  /mnt/
[root@xiandian ~]#
[root@xiandian ~]# mount -o loop XianDian-IaaS-v2.2.iso  /mnt/
mount: /dev/loop0 is write-protected, mounting read-only
[root@xiandian ~]# cp -rf /mnt/* /opt/
[root@xiandian ~]#
[root@xiandian ~]# umount  /mnt/
```

图 1-25　挂载情况

b. 配置 yum 路径。

将网络 yum 源路径移除 yum 目录。

```
[root@ compute ~]  # mv /etc/yum.repos.d/* /opt
```

执行没有返回结果。

c. 在计算节点创建 repo 文件。

在/etc/yum. repos. d 创建 centos. repo 源文件。

```
[root@ compute ~]  # vi /etc/yum.repos.d/local.repo
```

按【I】键进入编译模式,添加以下内容到 local. repo 文件里:

```
[centos]
name = centos
baseurl = file:///opt/centos
gpgcheck = 0
enabled = 1
[iaas]
name = iaas
baseurl = file:///opt/iaas - repo
gpgcheck = 0
enabled = 1
```

按【Esc】键并输入:wq 命令保存退出。

⑤配置 IP。

根据服务器自身的 IP 地址配置网卡信息。

```
[root@ compute ~]  # vi /etc/sysconfig/network - scripts/ifcfg - eno16777736
```

按【I】键进入编译模式,将文件修改为以下形式:

```
TYPE = Ethernet
BOOTPROTO = static
NM_CONTROLLED = yes
DEVICE = eno16777736
ONBOOT = yes
IPADDR = 192.168.100.20
PREFIX = 23
GATEWAY = 192.168.100.1
```

按【Esc】键并输入:wq 命令保存退出。

注:将获取 IP 方式改为静态。

在 IPADDR 输入自己虚拟机的 IP。

网关也是结合服务器所在网段填写。

```
[root@ compute ~]  # vi /etc/sysconfig/network - scripts/ifcfg - eno33554960
```

按【I】键进入编译模式,将文件修改为以下形式:

```
TYPE=Ethernet
BOOTPROTO=static
NM_CONTROLLED=yes
DEVICE=eno33554960
ONBOOT=yes
IPADDR=192.168.200.20
PREFIX=24
NETGATE=192.168.200.1
```

按【Esc】键并输入:wq 命令保存退出。

注:将获取 IP 方式改为静态。

在 IPADDR 输入自己虚拟机的 IP。

外网段不能和虚拟机 IP 在同一个网段。

重启网络:

```
[root@ compute ~] #systemctl restart network
```

⑥计算节点虚拟机安装 iaas - xiandian 安装包。

```
[root@ compute ~] #yum install -y iaas-xiandian -y
```

下载情况如图 1-26 所示。

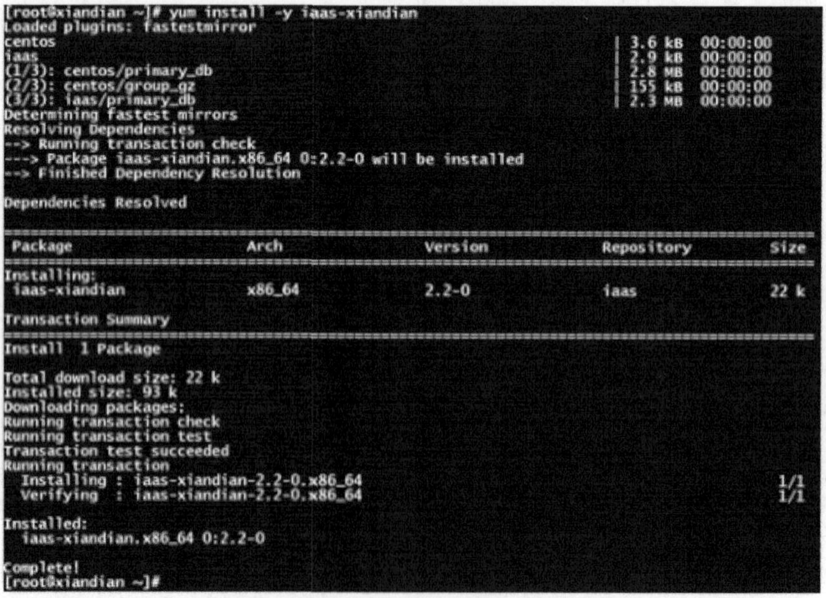

```
[root@xiandian ~]# yum install -y iaas-xiandian
Loaded plugins: fastestmirror
centos                                              | 3.6 kB  00:00:00
iaas                                                | 2.9 kB  00:00:00
(1/3): centos/primary_db                            | 2.8 MB  00:00:00
(2/3): centos/group_gz                              | 155 kB  00:00:00
(3/3): iaas/primary_db                              | 2.3 MB  00:00:00
Determining fastest mirrors
Resolving Dependencies
--> Running transaction check
---> Package iaas-xiandian.x86_64 0:2.2-0 will be installed
--> Finished Dependency Resolution

Dependencies Resolved

================================================================================
 Package            Arch           Version          Repository          Size
================================================================================
Installing:
 iaas-xiandian      x86_64         2.2-0            iaas                22 k

Transaction Summary
================================================================================
Install  1 Package

Total download size: 22 k
Installed size: 93 k
Downloading packages:
Running transaction check
Running transaction test
Transaction test succeeded
Running transaction
  Installing : iaas-xiandian-2.2-0.x86_64                                  1/1
  Verifying  : iaas-xiandian-2.2-0.x86_64                                  1/1

Installed:
  iaas-xiandian.x86_64 0:2.2-0

Complete!
[root@xiandian ~]#
```

图 1-26 下载情况

编辑文件/etc/xiandian/openrc.sh(配置环境变量)。

```
[root@ compute ~] #vi /etc/xiandian/openrc.sh
```

按【I】键进入编译模式，按以下参数修改配置文件。

说明：原配置文件中是有“#”号的，填写好配置后删除注释“#”。

```
HOST_IP =192.168.100.10
HOST_NAME = controller
HOST_IP_NODE =192.168.100.20
HOST_NAME_NODE = compute
RABBIT_USER = openstack
RABBIT_PASS =000000
DB_PASS =000000
DOMAIN_NAME = demo(自定义)
ADMIN_PASS =000000
DEMO_PASS =000000
KEYSTONE_DBPASS =000000
GLANCE_DBPASS =000000
GLANCE_PASS =000000
NOVA_DBPASS =000000
NOVA_PASS =000000
NEUTRON_DBPASS =000000
NEUTRON_PASS =000000
METADATA_SECRET =000000
INTERFACE_NAME = eno 33554960
CINDER_DBPASS =000000
CINDER_PASS =000000
BLOCK_DISK = vda3
SWIFT_PASS =000000
OBJECT_DISK = vda4
STORAGE_LOCAL_NET_IP =127.0.0.1
HEAT_DBPASS =000000
HEAT_PASS =000000
CEILOMETER_DBPASS =000000
CEILOMETER_PASS =000000
AODH_DBPASS =000000
AODH_PASS =000000
```

按【Esc】键并输入:wq 命令保存退出。

⑦在计算节点虚拟机上执行。

重启设备，重启会卡住需要单击关机并退出按钮，退出虚拟机，然后再次进入实验。

```
[root@ compute ~]  # iaas-pre-host.sh
```

```
[root@ compute  ~]  # reboot
```

注:如果没有反应,请重新刷新平台页面。

⑧部署脚本安装平台。

在计算节点虚拟机节点执行脚本 iaas - install - nova - compute. sh 进行 nova 计算服务安装。

```
[root@ compute ~]  #iaas-install-nova-compute.sh
```

在计算节点虚拟机执行脚本 iaas - install - neutron - compute. sh 进行 neutron 网络服务安装。

```
[root@ compute ~]  #iaas-install-neutron-compute.sh
```


在计算节点虚拟机执行脚本 iaas - install - neutron - compute - gre. sh 进行 gre 网络安装配置。

```
[root@ compute ~]  #iaas-install-neutron-compute-gre.sh
```

(8)验证安装基础工作完成。

上述操作完成后,打开网页 http://192. 168. 100. 10/dashboard 进行验证服务,图 1-27所示是 Dashboard 登录界面,包含管理员账号和密码登录。图 1-28 所示是 Dashboard 管理界面,表示安装基础工作正确完成。

注:这里填写自己虚拟机的 IP 地址进入 Dashboard。

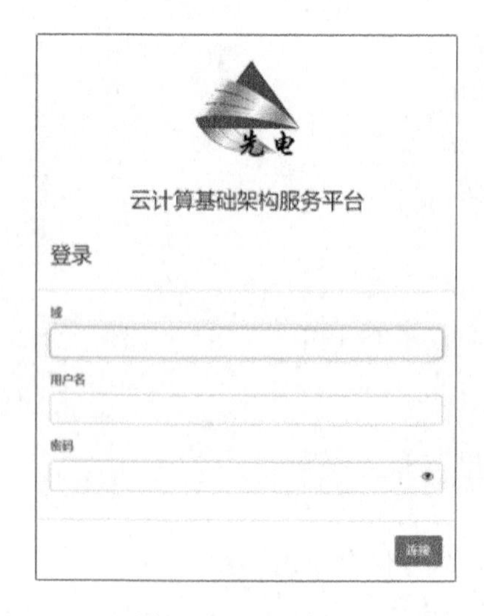

图 1-27　登录界面

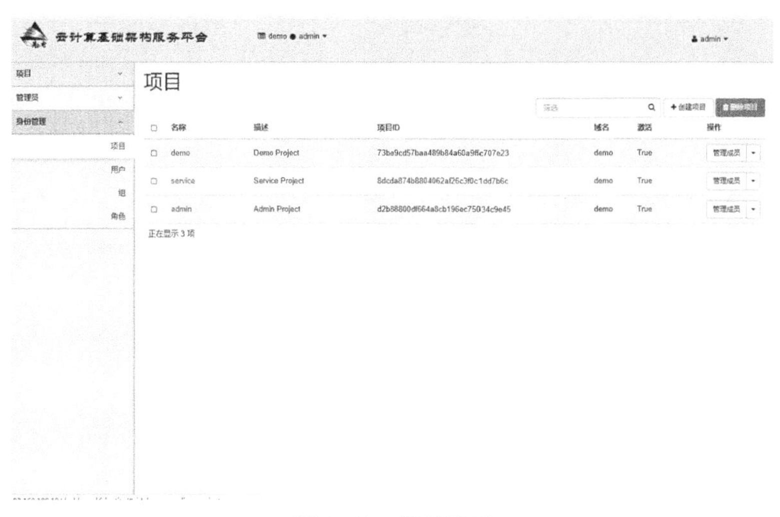

图 1-28　管理界面

(9)检测问题。

如果 Dashboard 界面不能访问,可通过另外一种方式检验。

```
[root@controller ~]# curl -L http://192.168.100.10/dashboard
```

返回结果是一个 HTML 网页,代码量较多,图 1-29 所示只截取了开始的一部分。

```
[root@controller ~]# curl -L http://192.168.100.10/dashboard

<!DOCTYPE html>
<html>
  <head>
    <meta content='IE=edge' http-equiv='X-UA-Compatible' />
    <meta content='text/html; charset=utf-8' http-equiv='Content-Type' /
>

    <title>Login - XianDian Dashboard</title>


<link rel="stylesheet" href="/dashboard/static/dashboard/css/34f8a8f8d5e
5.css" type="text/css" media="screen" />



<link rel="stylesheet" href="/dashboard/static/dashboard/css/e673225ff3b
e.css" type="text/css" />

<link rel="shortcut icon" href="/dashboard/static/dashboard/img/favicon.
ico"/>


  <style id="anti-clickjack">body{display:none !important;}</style>

  <script type='text/javascript' charset="utf-8">
    if (self === top) {
      var antiClickjack = document.getElementById("anti-clickjack");
      antiClickjack.parentNode.removeChild(antiClickjack);
    } else {
      top.location = self.location;
    }
  </script>



<script type="text/javascript" src="/dashboard/static/dashboard/js/7fd3f
7d69c71.js"></script>

<script type='text/javascript' charset='utf-8'>
  (function (global) {
    'use strict';

    // make translation info available on client
    horizon.languageCode = 'en';
    horizon.languageBidi = 'False';
    horizon.datepickerLocale = 'en';

    /* Load angular modules extensions list before we include angular/ho
rizon.js */
    global.horizonPlugInModules = ['horizon.dashboard.project', 'horizon
.dashboard.admin', 'horizon.dashboard.identity'];
```

图 1-29　返回结果

项目小结

本项目是云计算基础构建的开篇，主要介绍云计算起源及发展历程及根据企业的需求，使用 OpenStack 开源软件为企业设计云平台系统架构和安装云平台的基础工作。

项目扩展

本项目 Xiandian－IaaS－2.2 镜像包是来自南京第五十五所技术开发有限公司的先电品牌 CLOUD－TR200 型号的产品之一，它是基于以 Apache 开放许可证授权开源云计算项目 OpenStack 开发的。其实安装 OpenStack 还有多种方式，从源代码构建到用包安装都可以，但是项目一的方法是最简单和最常用的。

项目思考

学习完项目一的方法构建云平台，请思考还有哪些方式可以使用 OpenStack 来安装搭建云平台。

项目二 认证服务

项目综述

小张了解云计算的基础概念及搭建云计算平台的相关知识，在此基础上已经完成了公司云计算平台的系统环境设计和系统搭建的基础安装工作。在公司部署好的云计算平台下，要学习为公司员工创建用户账号、管理用户权限。具体任务如下：

视频
项目二综述

- 使用 Keystone 管理账号的认证服务。
- 创建租户、用户并绑定用户权限。

项目目标

【知识目标】

- OpenStack 的 Keystone 服务组件的相关概念。
- OpenStack 的 Keystone 服务申请认证机制流程。

【技能目标】

- 配置 Keystone 应用环境。
- 管理认证用户。
- 创建租户、用户并绑定用户权限。

【职业能力】

能使用 Keystone 组件来管理云平台的用户、租户及修改用户权限。

任务一　安装与配置 Keystone 认证服务

任务要求

小张在公司部署好的云计算平台下，学习如何配置 Keystone 认证服务，并学习为公司员工创建用户账号、管理用户权限。具体要求如下：

（1）配置并启用认证服务。

（2）创建项目 test。

（3）创建用户账号 alice。

（4）创建租户 acme，用于管理一组账户。

（5）创建角色 compute－user，用于用户权限的管理。

（6）绑定用户和租户的权限。

核心概念

认证服务：在 OpenStack 框架中，认证服务模块 Keystone 的功能是负责校验服务规则和发布服务令牌的，它实现了 OpenStack 的 Identity API。Keystone 可分解两个功能，即权限管理和服务目录。权限管理主要用于用户的管理授权。服务目录，类似于一个服务总线，或者说是整个 OpenStack 框架的注册表。认证模块提供 API 服务、Token 令牌机制、服务目录、规则和认证发布等功能。

知识准备

视频

认证服务的基本概念

1．相关概念

作为 OpenStack 的基础支持服务的 Keystone ，它有以下的功能：

①管理用户及其权限。

②维护 OpenStack Services 的 Endpoint（服务端点）。

③Authentication（认证）和 Authorization（鉴权）。

学习 Keystone，需要理解下面的概念：

（1）用户（User）。User 指任何使用 OpenStack 的实体，可以是真正的用户，其他系统或者服务或系统使用 OpenStack 相关服务的一个组织。当 User 请求访问 OpenStack 时，Keystone 会对其进行验证。

（2）证书（Credentials）。Credentials 是 User 用来证明自己身份的信息，可以是用户名/密码或 Token 或 API Key 或其他高级方式。

（3）认证（Authentication ）。Authentication 是 Keystone 验证 User 身份的过程。User 访问 OpenStack 时向 Keystone 提交用户名和密码形式的 Credentials，Keystone 验证通过后会给 User 签发一个 Token 作为后续访问的 Credential。用户在随后的请求中使用这个令牌去访问资源中的其他应用。

（4）令牌（Token）。Token 是由数字和字母组成的字符串，User 成功认证后 Keystone 生成 Token 并分配给 User。Token 用作访问 Service 的 Credential，Service 会通过 Keystone 验证 Token 的有效性，而且 Token 的有效期默认是 24 小时。

（5）租户（Group）。Group 即组，它是各个服务中的一些可以访问的资源集合。比如通过 Nova 创建虚拟机时要指定到某个组中，在 cinder 创建卷也要指定到某个组中，用户访问租户的资源前，必须与该组关联，并且指定该用户在该组下的角色。

（6）工程（Project）。Project 即工程，用于将 OpenStack 的资源（计算、存储和网络）进行分组和隔离。

根据 OpenStack 服务的对象不同，Project 可以是一个客户（公有云，又称租户）、部门或者项目组（私有云）。资源的所有权是属于 Project 的，而不是 User。在 OpenStack 的界面和文档中，Group、Project、Account 这几个术语是通用的，但长期看会倾向使用 Project。每个 User（包括 admin）必须挂在 Project 里才能访问该 Project 的资源。一个 User 可以属于多个 Project。Admin 相当于 root 用户，具有最高权限。

（7）角色（Role）。Role 即角色，Role 代表一组用户可以访问的资源权限，如 Nova 中的虚拟机、Glance 中的镜像。Users 可以被添加到任意一个全局的 Role 或租户内的 Role 中。在全局的 Role 中，用户的 Role 权限作用于所有的租户，即可以对所有的租户执行 Role 规定的权限，在租户内的 Role 中，用户仅能在当前租户内执行 Role 规定的权限。

2. 认证服务流程

假设用户 tony 请求云主机的流程涉及认证 Keystone 服务，计算 Nova 服务、镜像 Glance 服务，在服务流程中，令牌（Token）作为流程认证传递，具体服务申请认证机制流程如图 2－1 所示。

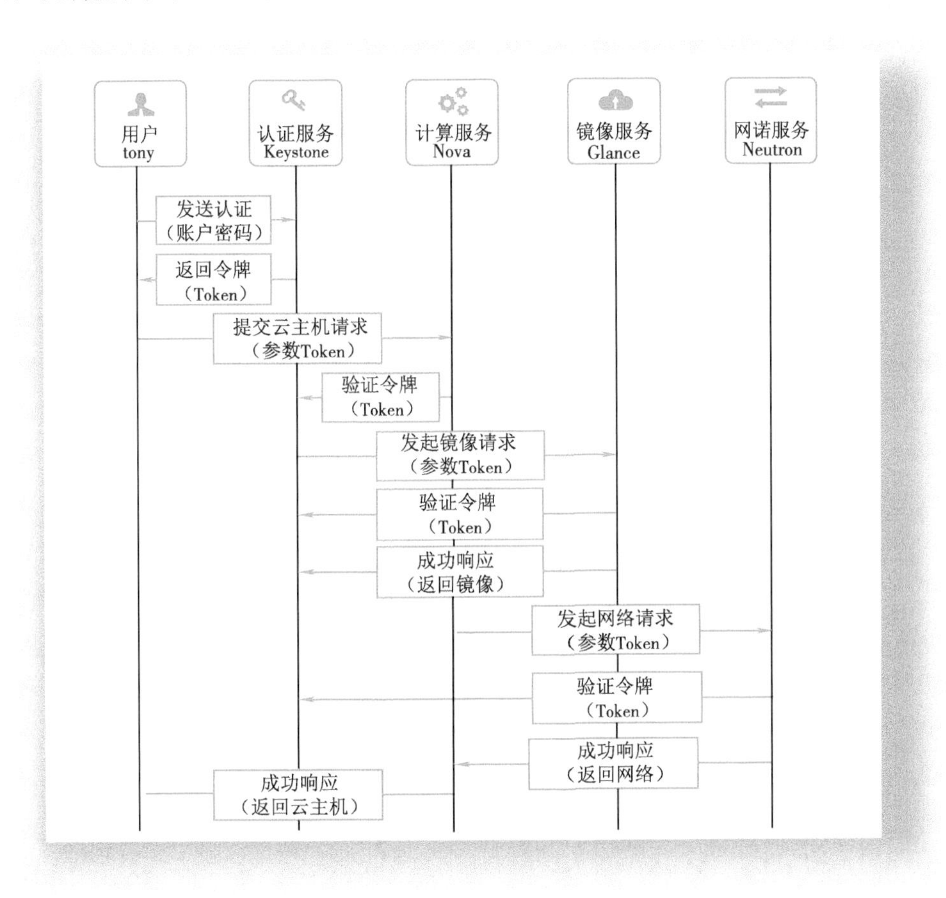

图 2－1　Keystone 的认证流程

1. 准备实施任务的环境（见表 2－1）

表 2－1　实施任务的准备环境

实施任务所需软件资源	虚拟机镜像资源信息
虚拟机软件（vm12）	controller
securcrt 远程连接软件	compute

2. 分解实施任务的过程（见表 2－2）

表 2－2　实施任务的简明过程

序号	步骤	详细操作及说明
1	配置 Keystone 应用环境	
2	管理认证用户	①创建用户； ②创建项目； ③创建角色； ④绑定用户和项目权限

3. 配置 Keystone 应用环境

在安装 Keystone 服务之前需要指定用户名和密码，通过认证服务来进行身份认证，在开始阶段是没有创建任何用户的。因此必须用授权令牌的服务的访问接口来创建特定的用来进行身份认证的用户，之后需要创建一个管理用户的环境变量(admin－openrc. sh)来管理最终的凭证和终端。

在安装 Keystone 服务之后，产生的主配置文件存放在/etc/keystone 目录中，名称为 keystone. conf，在配置文件中需要配置初始的 Token 值和数据库的连接地址。

Keystone 服务安装完毕，可以通过请求身份令牌来验证服务，具体命令如下(以 admin 用户访问 http://controller:35357/v3 地址获取 Token 值)：

```
[root@ controller ~] #openstack --os-project-name admin --os-domain-name demo --os-username admin --os-password 000000 --os-auth-url http://controller:35357/v3 token issue
```

注：如执行错误，请等待 2～3 秒后重新执行该命令。

执行返回结果如图 2－2 所示。

图 2－2　admin 用户访问地址获取 Token 值

4. 管理认证用户

OpenStack 的用户（User）包括云平台使用者、服务以及系统。用户通过认证登录系统并调用资源。为方便被分配到一个或多个租户，租户是用户的集合。为给用户分配不同的权限，Keystone 设置了角色（Role），角色是代表用户可以访问的资源等权限。用户可以被添加到任意一个全局的或租户内的角色中。在全局的角色中，用户的角色权限作用于所有的用户，即可以对所有的用户执行角色规定的权限；租户内的角色，用户仅能在当前的租户内执行角色规定的权限，下面的操作将实现以下的任务要求：

①创建用户。

②创建项目。

③创建角色。

④绑定用户和租户的权限。

具体代码如下：

（1）创建用户。在 OpenStack 系统中进行操作需生效环境变量，执行命令如下：

```
[root@ controller ~] # source /etc /keystone /admin -
openrc. sh
```

创建一个名称为 alice 的账户，密码为 mypassword123，邮箱为 alice@ example. com。执行命令如下：

```
[root @ controller ~] # openstack user create --password
mypassword123 --email alice@ example. com --domain demo alice
```

执行返回结果如图 2-3 所示。

```
[root@controller ~]# openstack user create --password mypassword123 --email alice@example.com --domain demo alice
+-----------+----------------------------------+
| Field     | Value                            |
+-----------+----------------------------------+
| domain_id | b4d22478e683415c88cb019e2abaf8e8 |
| email     | alice@example.com                |
| enabled   | True                             |
| id        | 7820759de3fb4e3cacfde5a6dee351d9 |
| name      | alice                            |
```

图 2-3　创建 alice 账户

从上面的操作可以看出，创建用户需要用户名称、密码和邮件等信息。具体格式如下：

```
# openstack user create [ --domain  <domain >]
[ --password  <password >]
[ --email  <email - address >]
[ --enable | --disable]
<name >
```

其中，参数 <name >代表新建用户名。

（2）创建项目。一个项目就是一个项目、团队或组织，当请求 OpenStack 服务时，必须定义一个项目。例如，查询计算服务正在运行的云主机实例列表。

创建一个名为 acme 的项目。执行命令如下：

```
[root@ controller ~] # openstack project create --domain
demo acme
```

执行返回结果如图 2-4 所示。

```
[root@xiandian ~]# openstack project create --domain xiandian acme
+-------------+----------------------------------+
| Field       | Value                            |
+-------------+----------------------------------+
| description |                                  |
| domain_id   | 3ac89594c8e944a9b5bb567fca4e75aa |
| enabled     | True                             |
| id          | 2e1647a7f1c94acb9ac9aff1b91ad5ad |
| is_domain   | False                            |
| name        | acme                             |
| parent_id   | 3ac89594c8e944a9b5bb567fca4e75aa |
+-------------+----------------------------------+
```

图 2-4　创建 acme 项目

从上面的操作可以看出，创建项目需要项目名等相关信息。具体操作格式如下：

```
# openstack project create [ --domain <domain > ]
[ --description <description >]
[ --enable | --disable]
<project - name >
```

其中，参数 <project - name> 代表新建项目名，参数 <description> 代表项目描述名。

（3）创建角色。角色限定了用户的操作权限。例如，创建一个角色 compute - user，执行命令如下：

```
[root@ controller ~] # openstack role create compute - user
```

执行返回结果如图 2-5 所示。

```
[root@xiandian ~]# openstack role create compute-user
+-----------+----------------------------------+
| Field     | Value                            |
+-----------+----------------------------------+
| domain_id | None                             |
| id        | e65851d697844aafbbe7c122cf7e88b7 |
| name      | compute-user                     |
+-----------+----------------------------------+
```

图 2-5　创建 compute - user 角色

从上面的操作可以看出，创建角色需要角色名称信息。具体命令格式如下：

```
# openstack role create <name >
```

其中，参数 <name> 代表角色名称。

（4）绑定用户和项目权限。

添加的用户需要分配一定的权限，这就需要把用户关联绑定到对应的项目和角色。例如，给用户 alice 分配 acme 项目下的 compute - user 角色，执行命令如下：

```
[root@ controller ~] # openstack role add --user alice --project acme compute - user
```

从上面的操作可以看出，绑定用户权限需要用户名称、角色名称和项目名称等信息。具体命令格式如下：

```
# openstack role add --user <name> --project <project> <role>
```

其中，参数 <name> 代表需要绑定的用户名称，参数 <role> 代表用户绑定的角色名称，参数 <project> 代表用户绑定的项目名称。

任务二　创建租户、用户并绑定用户权限

任务要求

小张经过一系列学习之后，已经初步掌握到 Keystone 认证服务的使用方法，现在小张要为公司的员工创建相应的部门租户，为员工创建员工用户，并赋予相应的权限。具体要求如下：

公司有 10 名员工，其中 2 名为项目研发部（研发环境），3 名为业务部（办公环境），5 人 IT 工程部（运维环境）。根据企业人员部门分配，现构建 3 个租户，10 个用户，管理人员拥有管理员权限，其余人员拥有普通用户权限，具体如表 2 - 3 所示。

表 2 - 3　用户权限

部门	租户	用户	权限
项目研发部	RD_ Dept	rduser 01/rduser 02	普通用户
业务部	BS_ Dept	bsuser 01 ~ bsuser 03	普通用户和管理员用户
IT 工程部	IT_ Dept	ituser 01 ~ ituser 05	普通用户

知识准备

OpentStack 服务（Service），如 Keystone、Nova、Glance、Swift、Heat、Ceilometer 等。为了方便用户调用这些服务，OpenStack 为每一个服务提供一个用于访问的端点（Endpoit），如果需要访问服务，则必须知道它的端点，端点一般为 URL。如果知道服务的 URL，就可以访问它。端点的 URL 具有 public、private 和 admin 三种权限。Public URL 可以被全局访问，private URL 只能被局域网访问，admin URL 从常规的访问中被分离出来。

常用的服务管理命令如下：

1. 创建服务

具体功能：创建服务，如图 2－6 所示。

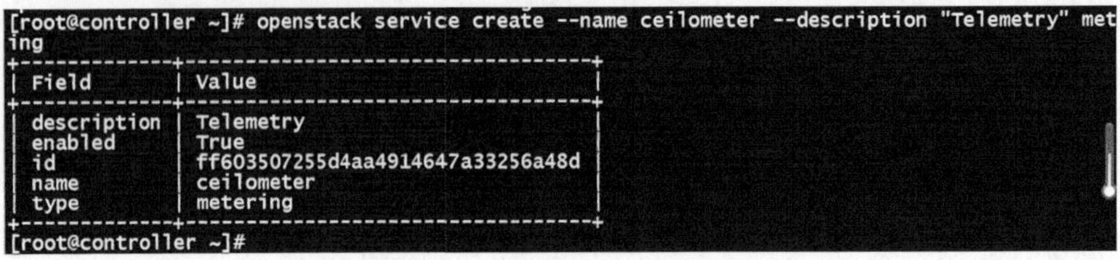

```
[root@controller ~]# openstack service create --name ceilometer --description "Telemetry" metering
+-------------+----------------------------------+
| Field       | Value                            |
+-------------+----------------------------------+
| description | Telemetry                        |
| enabled     | True                             |
| id          | ff603507255d4aa4914647a33256a48d |
| name        | ceilometer                       |
| type        | metering                         |
+-------------+----------------------------------+
[root@controller ~]#
```

图 2－6　创建服务

命令格式如下：

```
#openstack service create --name <name> --type <type>
    [--description <service-description>]
      --name <name>创建服务名称
      --type <type>创建服务类型
      --description <service-description>创建服务描述
```

2. 创建服务访问端点

具体功能：创建服务访问的 API 端点，如图 2－7 所示。

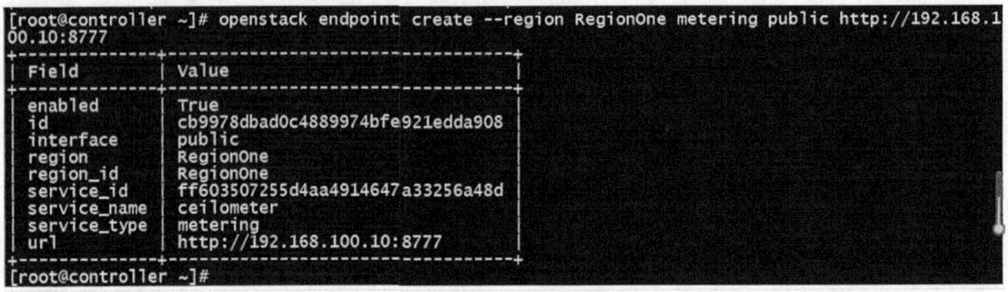

```
[root@controller ~]# openstack endpoint create --region RegionOne metering public http://192.168.100.10:8777
+--------------+----------------------------------+
| Field        | Value                            |
+--------------+----------------------------------+
| enabled      | True                             |
| id           | cb9978dbad0c4889974bfe921edda908 |
| interface    | public                           |
| region       | RegionOne                        |
| region_id    | RegionOne                        |
| service_id   | ff603507255d4aa4914647a33256a48d |
| service_name | ceilometer                       |
| service_type | metering                         |
| url          | http://192.168.100.10:8777       |
+--------------+----------------------------------+
[root@controller ~]#
```

图 2－7　创建服务访问端点

命令格式如下：

```
#openstack endpoint create [--region <endpoint-region>]
--service <service> --publicurl <public-url>
        [--adminurl <admin-url>]
        [--internalurl <internal-url>]
      --region <endpoint-region>创建端点的区域名称
      --service <service>端点创建的使用服务名称
      --publicurl <public-url>对外服务的 URL 地址
      --adminurl <admin-url>管理网络访问的 URL 地址
      --internalurl <internal-url>内部访问的 URL 地址
```

3. 查询服务目录

Service Catalog（服务目录）是 Keystone 为 OpenStack 提供的一个 REST API 端点列表，并以此作为决策参考。

#openstack catalog list #可以显示所有已有的 service

#openstack catalog list --service <service-type> #显示某个 service 信息

查询服务目录如图 2-8 所示。

```
[root@controller ~]# openstack catalog list
+-----------+----------------+----------------------------------------------------------------+
| Name      | Type           | Endpoints                                                      |
+-----------+----------------+----------------------------------------------------------------+
| swift     | object-store   | RegionOne                                                      |
|           |                |   admin: http://controller:8080/v1                             |
|           |                | RegionOne                                                      |
|           |                |   internal: http://controller:8080/v1/AUTH_26b08dcad5a74c0     |
|           |                | 9b1ad0fda0a611bce                                              |
|           |                | RegionOne                                                      |
|           |                |   public: http://controller:8080/v1/AUTH_26b08dcad5a74c09b     |
|           |                | 1ad0fda0a611bce                                                |
|           |                |                                                                |
| nova      | compute        | RegionOne                                                      |
|           |                |   public: http://controller:8774/v2.1/26b08dcad5a74c09b1ad     |
|           |                | 0fda0a611bce                                                   |
|           |                | RegionOne                                                      |
|           |                |   internal: http://controller:8774/v2.1/26b08dcad5a74c09b1     |
|           |                | ad0fda0a611bce                                                 |
|           |                | RegionOne                                                      |
|           |                |   admin: http://controller:8774/v2.1/26b08dcad5a74c09b1ad0     |
|           |                | fda0a611bce                                                    |
|           |                |                                                                |
| cinder    | volume         | RegionOne                                                      |
|           |                |   admin:                                                       |
|           |                | http://controller:8776/v1/26b08dcad5a74c09b1ad0fda0a611bce     |
|           |                | RegionOne                                                      |
|           |                |   internal:                                                    |
|           |                | http://controller:8776/v1/26b08dcad5a74c09b1ad0fda0a611bce     |
|           |                | RegionOne                                                      |
|           |                |   public:                                                      |
|           |                | http://controller:8776/v1/26b08dcad5a74c09b1ad0fda0a611bce     |
|           |                |                                                                |
| cinderv2  | volumev2       | RegionOne                                                      |
|           |                |   internal:                                                    |
|           |                | http://controller:8776/v2/26b08dcad5a74c09b1ad0fda0a611bce     |
|           |                | RegionOne                                                      |
|           |                |   public:                                                      |
|           |                | http://controller:8776/v2/26b08dcad5a74c09b1ad0fda0a611bce     |
|           |                | RegionOne                                                      |
|           |                |   admin:                                                       |
|           |                | http://controller:8776/v2/26b08dcad5a74c09b1ad0fda0a611bce     |
|           |                |                                                                |
| heat-cfn  | cloudformation | RegionOne                                                      |
|           |                |   internal: http://controller:8000/v1                          |
```

图 2-8 查询服务目录

4. 查询 Keystone 服务和授权协议

#openstack catalog show keystone

查询 Keystone 服务和授权协议，如图 2-9 所示。

```
[root@controller ~]# openstack catalog show keystone
+-----------+--------------------------------------------+
| Field     | Value                                      |
+-----------+--------------------------------------------+
| endpoints | RegionOne                                  |
|           |   public: http://controller:5000/v3        |
|           | RegionOne                                  |
|           |   admin: http://controller:35357/v3        |
|           | RegionOne                                  |
|           |   internal: http://controller:5000/v3      |
|           |                                            |
| id        | 83a264ee025e4570bbbb41c49a58d051           |
| name      | keystone                                   |
| type      | identity                                   |
+-----------+--------------------------------------------+
[root@controller ~]#
```

图 2-9 查询 Keystone 服务和授权协议

创建租户、用户并绑定用户权限

1. 准备实施任务的环境（见表 2－4）

表 2－4　实施任务的准备环境

实施任务所需软件资源	虚拟机镜像资源信息
虚拟机软件（vm12）	controller
securcrt 远程连接软件	compute

2. 分解实施任务的过程（见表 2－5）

表 2－5　实施任务的简明过程

序号	步骤	详细操作及说明
1	创建租户	Dashboard 界面与 CLI 命令创建
2	创建用户账号	CLI 命令创建
3	绑定用户权限	Dashboard 界面与 CLI 命令创建

3. 创建租户（即是工程 project）

创建项目研发部门名为 RD_Dept 的租户、业务部门名为 BS_Dept 的租户和 IT 工程部门名为 IT_Dept 的租户，如图 2－10 所示。

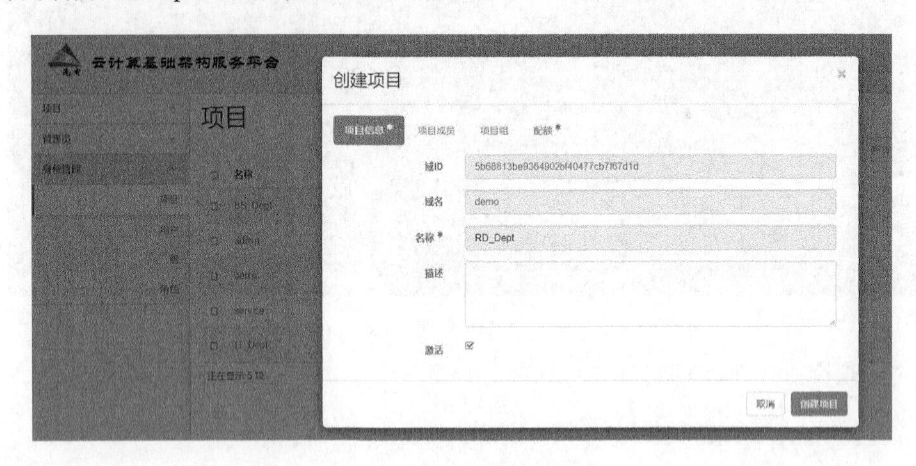

图 2－10　创建租户 1

（1）通过 Dashboard 界面为项目研发部门创建一个名为 RD_Dept 的租户，步骤如下：

①进入 Dashboard 找到管理员选项。

②选择“创建项目”，在弹出窗口中输入名称和项目描述信息，默认情况下，项目是自动激活的。

③在配额选项中，进行项目资源分配。

创建成功后在项目列表中，会显示出该项目条目，并获得一个自动分配的 ID，如图 2－11 所示。

图 2－11　创建租户 2

（2）通过 CLI 界面为业务部门创建一个名为 BS_ Dept 的租户，如图 2－12 所示。

[root@ controller ~] # openstack project create　--domain = demo BS_ Dept

```
[root@controller ~]# openstack project create  --domain=demo   BS_Dept
+-------------+----------------------------------+
| Field       | Value                            |
+-------------+----------------------------------+
| description |                                  |
| domain_id   | 5b68813be9364902bf40477cb7f67d1d |
| enabled     | True                             |
| id          | 04d66e6429584c4abeba3e6c2465b441 |
| is_domain   | False                            |
| name        | BS_Dept                          |
| parent_id   | 5b68813be9364902bf40477cb7f67d1d |
+-------------+----------------------------------+
[root@controller ~]#
```

图 2－12　创建租户 3

（3）通过 CLI 界面为 IT 工程部创建一个为 IT_ Dept 的租户，如图 2－13 所示。

[root@ controller ~] # openstack project create　--domain = demo IT_ Dept

```
[root@controller ~]# openstack project create  --domain=demo   IT_Dept
+-------------+----------------------------------+
| Field       | Value                            |
+-------------+----------------------------------+
| description |                                  |
| domain_id   | 5b68813be9364902bf40477cb7f67d1d |
| enabled     | True                             |
| id          | ffd401400c2e4fd581a5d3d9aa962d5a |
| is_domain   | False                            |
| name        | IT_Dept                          |
| parent_id   | 5b68813be9364902bf40477cb7f67d1d |
+-------------+----------------------------------+
[root@controller ~]#
```

图 2－13　创建租户 4

4. 创建用户账号

为项目研发部创建 2 个用户，分别名为 rduser 01 和 rduser 02，密码为 cloudpasswd；为业务部创建 3 个用户，分别名为 bsuser 01 ~ bsuser 03，密码为 cloudpasswd；为 IT 工程部门创建 5 个用户，分别名为 ituser 01 ~ ituser 05，密码为 cloudpasswd。可使用 Dashboard 和 CLI 界面，可辅助使用 shell。

（1）为项目研发部创建 2 个用户，分别名为 rduser 01 和 rduser 02，密码为 cloudpasswd。

```
[root@ controller ~] #openstack user create --domain=demo
--password=cloudpasswd --email=alice@ example. com --
project=RD_ Dept  rduser01
```

其他用户类似，如图 2－14 和图 2－15 所示。

（2）为业务部创建 3 个用户，分别名为 bsuser 01 ~ bsuser 03，密码为 cloudpasswd。

```
[root@ controller ~] #openstack user create --domain=demo
--password=cloudpasswd --email=alice@ example. com --
project=BS_ Dept  bsuser01
```

其他用户类似，如图 2－16 所示。

（3）为 IT 工程部门创建 5 个用户，分别名为 ituser 01 ~ ituser 05，密码为 cloudpasswd。

```
[root@ controller ~] #openstack user create --domain=demo
--password=cloudpasswd --email=alice@ example. com --
project=IT_ Dept  ituser01
```

其他用户类似，如图 2－17 所示。

```
[root@controller ~]# openstack user create --domain=demo --password=cloudpasswd--email=alice@example.com --project=RD_Dept  rduser01
+---------------------+----------------------------------+
| Field               | Value                            |
+---------------------+----------------------------------+
| default_project_id  | 58e6935e61184bc6b9bbc8f2275b6b42 |
| domain_id           | 5b68813be9364902bf40477cb7f67d1d |
| enabled             | True                             |
| id                  | d8d75432c0fa4c8fa3b0d7dbff6f305a |
| name                | rduser01                         |
+---------------------+----------------------------------+
[root@controller ~]#
```

图 2－14　创建项目研发部用户账号 1

```
[root@controller ~]# openstack user create --domain=demo --password=cloudpasswd--email=alice@example.com --project=RD_Dept  rduser02
+---------------------+----------------------------------+
| Field               | Value                            |
+---------------------+----------------------------------+
| default_project_id  | 58e6935e61184bc6b9bbc8f2275b6b42 |
| domain_id           | 5b68813be9364902bf40477cb7f67d1d |
| enabled             | True                             |
| id                  | 01e6806639fe4c8789fb58c22ba05a02 |
| name                | rduser02                         |
+---------------------+----------------------------------+
[root@controller ~]#
```

图 2－15　创建项目研发部用户账号 2

```
[root@controller ~]# openstack user create --domain=demo --password=cloudpasswd--email=alice@exa
mple.com  --project=BS_Dept  bsuser01
+---------------------+----------------------------------+
| Field               | Value                            |
+---------------------+----------------------------------+
| default_project_id  | 04d66e6429584c4abeba3e6c2465b441 |
| domain_id           | 5b68813be9364902bf40477cb7f67d1d |
| enabled             | True                             |
| id                  | b5db36f8d2fd4b3e8f64e2e2c498c6c4 |
| name                | bsuser01                         |
+---------------------+----------------------------------+
[root@controller ~]#
```

图 2－16　创建业务部用户账号

```
[root@controller ~]# openstack user create --domain=demo --password=cloudpasswd--email=alice@exa
mple.com  --project=IT_Dept  ituser01
+---------------------+----------------------------------+
| Field               | Value                            |
+---------------------+----------------------------------+
| default_project_id  | ffd401400c2e4fd581a5d3d9aa962d5a |
| domain_id           | 5b68813be9364902bf40477cb7f67d1d |
| enabled             | True                             |
| id                  | 9f002eadce7d4818be92cc4a70d4f597 |
| name                | ituser01                         |
+---------------------+----------------------------------+
[root@controller ~]#
```

图 2－17　创建 IT 工程部用户账号

5. 绑定用户权限

将项目研发部、业务部的用户绑定普通用户权限；将 IT 工程部的用户绑定管理员和普通用户权限。

（1）通过 Dashboard 界面将项目研发部用户 rduser 01 绑定普通用户权限。

①进入 Dashboard 界面找到管理员选项。

②打开认证面板，选中“项目”选项卡。

③找到相应的项目，在“操作”栏目中选择“管理用户”选项，进入“编辑项目”对话框。

④在“项目成员”区域中，为项目用户选择相应的角色，如图 2－18 所示。

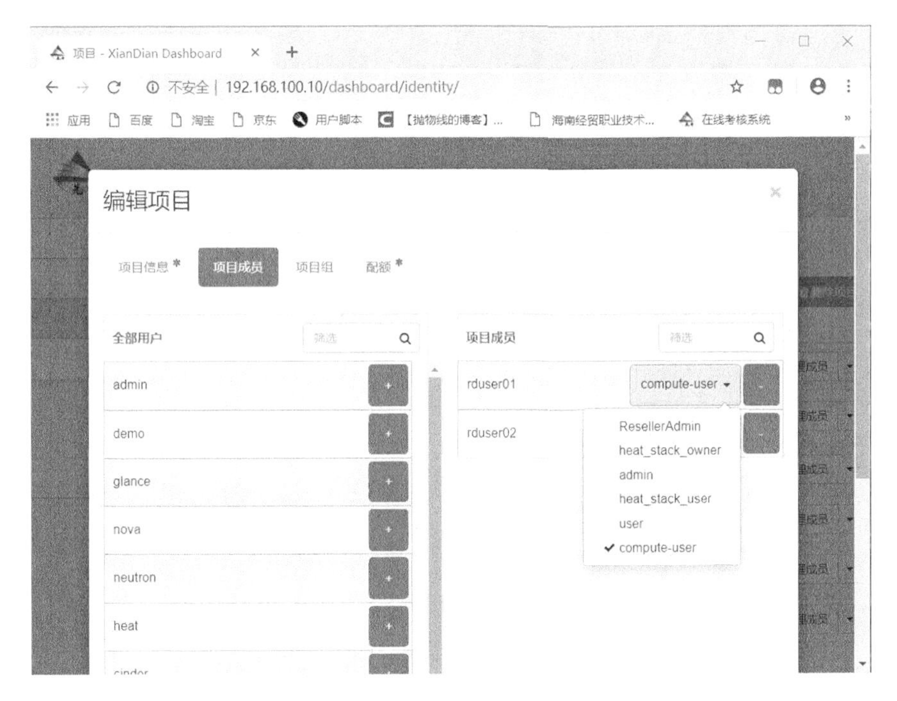

图 2－18　绑定用户权限 1

（2）通过 Shell 命令行将业务部用户 bsuser 01 绑定普通用户权限，其他的业务部用户类似，如图 2-19 所示。

```
[root@ controller ~] # openstack role add  --domain = demo
--user = bsuser01 compute - user
```

```
[root@controller ~]# openstack role add  --domain=demo  --user=rduser01 compute-user
[root@controller ~]# openstack role add  --domain=demo  --group=BS_Dept compute-user
[root@controller ~]#
```

图 2-19　绑定用户权限 2

（3）通过 Dashboard 界面将 IT 工程部用户 ituser 01 ~ ituser 05 绑定普通用户权限和管理员用户，如图 2-20 所示。

图 2-20　用户权限 3

项目小结

本项目主要讲解 OpenStack 的主要认证组件 Keystone，在 OpenStack 框架中，认证服务模块 Keystone 的功能是负责校验服务规则和发布服务令牌的，它实现了 OpenStack 的 Identity API，它是使用 OpenStack 服务的必经之路。同时也讲解 Keystone 的用户、证书、认证、令牌、租户、工程、角色的相关概念，并描述认证服务的过程。接下来，通过 Keystone 管理认证任务和创建租户、用户，并绑定用户权限两个任务使读者较好地完成项目目标。

项目扩展

表 2-6 列举本项目的组、用户、角色、工程和令牌的常用命令。

表 2－6　常用命令及其说明

常用命令	命令说明
openstack group add user	添加用户到组中
openstack group contains user	审核用户是否在组中
openstack group create	新增组
openstack group delete	删除组
openstack group list	列出组
openstack group remove user	移除用户出组
openstack group set	设置组内容
openstack group show	显示组内容
openstack project create	新增一个工程
openstack project delete	删除一个工程
openstack project list	列举工程内容
openstack project set	设置工程内容
openstack project show	显示工程内容
openstack role add	新增角色分配列表
openstack role assignment list	角色分配列表
openstack role create	新增角色
openstack role delete	删除角色
openstack role list	列举角色
openstack role remove	移除角色
openstack role set	设置角色内容
openstack role show	显示角色内容
openstack token issue	显示令牌
openstack token revoke	撤销现有令牌
openstack user create	新增用户
openstack user delete	删除用户
openstack user list	列出用户
openstack user password set	设置用户密码
openstack user set	设置用户信息
openstack user show	展示用户信息

结合本项目，假定用户 admin 要查看工程中的 image，OpenStack 内部 Keystone 发生了哪些事情？

项目三 消息队列服务

项目综述

小张在公司部署好的云计算平台下为公司员工创建用户账号、设置管理用户权限后。现在要了解 OpenStack 中的消息服务的基本状况和使用的情景，具体任务如下：掌握消息队列的基本操作及常见运维。

项目三综述

项目目标

【知识目标】

- OpenStack 的消息队列的相关概念。
- OpenStack 的 RabbitMQ 的服务流程和运行机制。

【技能目标】

掌握消息队列的基本操作及常见运维。

【职业能力】

了解 RabbitMQ 组件如何为云平台的正常运行提供支撑。

任务　安装与运行消息队列服务

任务要求

在日常的工作生活中，消息传递是一个必不可少的需求。在大型软件的内部信息交换和外部消息传递中，消息传递都是不可或缺的。在系统间通信窗体的最基本方法是 socket，但是这是一个底层的协议，在使用时需要程序来调用。在后序的学习任务中，小张需要先了解消息服务的基本状况和使用的情景，以及 OpenStack 的 RabbitMQ 的运行机制。

核心概念

消息服务：在 OpenStack 框架中，消息服务模块 RabbitMQ 的功能是帮助各个模块间完成消息传递。RabbitMQ 是通过队列、路由（包括点对点和发布 / 订阅）方式实现各个模块间的消息传递。

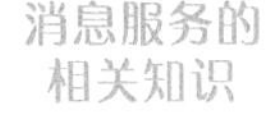

知识准备

1. 消息队列

消息队列（Message Queue，MQ），从字面意思上看，本质是个队列，FIFO 先入先出，只不过队列中存放的内容是 message 而已。其主要用途：不同进程 Process/线程 Thread 之间通信。为什么会产生消息队列？有以下几个原因：

（1）不同进程（Process）之间传递消息时，两个进程之间耦合程度过高，改动一个进程，引发必须修改另一个进程，为了隔离这两个进程，在两个进程间抽离出一层（一个模块），所有两个进程之间传递的消息，都必须通过消息队列来传递，单独修改某一个进程，不会影响另一个。

（2）不同进程（Process）之间传递消息时，为了实现标准化，将消息的格式规范化了，并且，某一个进程接收的消息太多，一下子无法处理完，也有先后顺序，必须对收到的消息进行排队，因此诞生了事实上的消息队列。

MQ 框架非常之多，比较流行的有 RabbitMQ、ActiveMQ、ZeroMQ、kafka，以及阿里开源的 RocketMQ。

2. RabbitMQ 消息服务

AMQP（Advanced Message Queuing Protocol，高级消息队列协议）是应用层协议的一个开放标准，为面向消息的中间件设计。消息中间件主要用于组件之间的解耦，消息的发送者无须知道消息使用者的存在，反之亦然。AMQP 的主要特征是面向消息、队列、路由（包括点对点和发布/订阅）、可靠性、安全。RabbitMQ 是一个开源的 AMQP 实现，服务器端用 Erlang 语言编写，支持多种客户端，如 Python、Ruby、. NET、Java、JMS、C、PHP、ActionScript、XMPP、STOMP 等，支持 AJAX。用于在分布式系统中存储转发消息，在易用性、扩展性、高可用性等方面表现不俗。

3. OpenStack 的 RabbitMQ

OpenStack 的 Nova 中各个组件之间的交互是通过“消息队列”来实现的，其中一种实现方法就是使用 RabbitMQ，有这样几个角色：producer、consumer、exchange、queue，其中 producer 是消息发送者，consumer 是消息接受者，中间要通过 exchange 和 queue。producer 将消息发送给 exchange，exchange 决定消息的路由，即决定要将消息发送给哪个 queue，然后 consumer 从 queue 中取出消息，进行处理，大致流程如图 3 -1 所示。

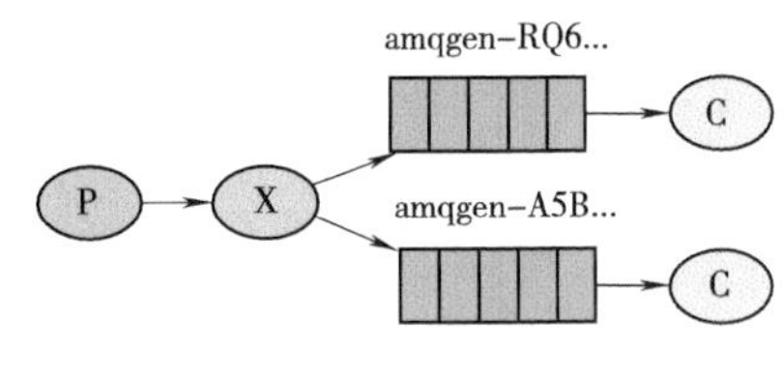

图 3 -1　消息流程

这几个角色中，最关键的是 exchange，有 3 种类型：direct、topic、fanout。其中，功能最强大的就是 topic，用它完全可以实现 direct 和 fanout 的功能。

（1）direct 是单条件的路由，即在 exchange 判断要将消息发送给哪个 queue 时，判断的依据只能是一个条件。

（2）fanout 是广播式的路由，即将消息发送给所有的 queue。

（3）topic 是多条件的路由，转发消息时，依据的条件是多个，所以只使用 topic 就可以实现 direct 和 fanout 的功能。

所说的"条件"反映到程序中，就是 routing_key，这个 routing_key 出现在两个地方：

· 每一个发送的消息都有一个 routing_key，表示发送的是一个什么样的消息。

· 每一个 queue 要和 exchange 绑定，绑定时要提供一个 routing_key，表示这个 queue 想要接收什么样的消息。

这样，exchange 就可以根据 routing_key，来将消息发送到合适的 queue 中。

视频 安装与运行消息队列RabbitMQ

1. 准备实施任务的环境（见表 3-1）

表 3-1 实施任务的准备环境

实施任务所需软件资源	虚拟机镜像资源信息
虚拟机软件（vm12）	controller
securcrt 远程连接软件	compute

2. 分解实施任务的过程（见表 3-2）

表 3-2 实施任务的简明过程

序号	步骤	详细操作及说明
1	检测 RabbitMQ 服务	
2	RabbitMQ 服务用户操作	①查询当前用户列表； ②创建 RabbitMQ 用户
3	赋予消息队列服务用户访问权限	①赋予 OpenStack 用户对所有资源读/写的权限； ②查询 OpenStack 用户所拥有的权限

3. 在 OpenStack 中安装与运行消息队列 RabbitMQ

（1）检测 RabbitMQ 服务。

检测当前环境是否已有 RabbitMQ 服务：

```
[root@ controller ~] # source /etc/keystone/admin-openrc.sh
[root@ controller ~] # rpm -qa |grep rabbitmq
```

结果如图 3-2 所示。

```
[root@xiandian ~]# rpm -qa |grep rabbitmq
rabbitmq-server-3.6.5-1.el7.noarch
```

图 3-2　检测服务

如果监测没有服务，就通过 yum install -y rabbitmq-server 进行安装。

（2）RabbitMQ 服务用户操作。

①查询当前用户列表。

```
[root@ controller ~] # rabbitmqctl list_ users
```

结果如 3-3 所示。

```
[root@xiandian ~]# rabbitmqctl list_users
Listing users ...
xiandian        []
guest   [administrator]
```

图 3-3　用户列表

②创建 RabbitMQ 用户。

```
[root@ controller ~] # rabbitmqctl add_ user openstack 000000
[root@ controller ~] # rabbitmqctl list_ users
```

结果如图 3-4 所示。

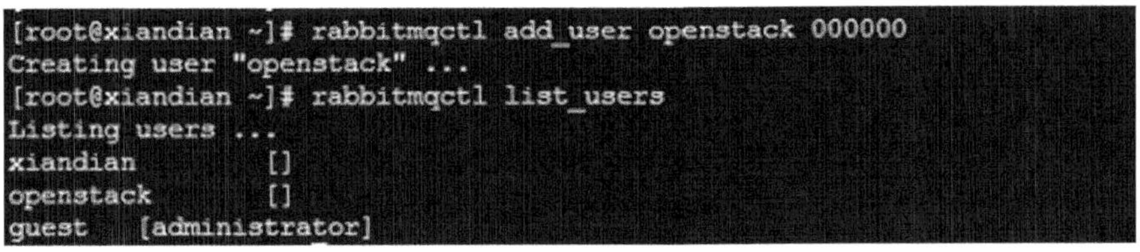

```
[root@xiandian ~]# rabbitmqctl add_user openstack 000000
Creating user "openstack" ...
[root@xiandian ~]# rabbitmqctl list_users
Listing users ...
xiandian        []
openstack       []
guest   [administrator]
```

图 3-4　创建用户

（3）赋予消息队列服务用户访问权限。

①赋予 OpenStack 用户对所有资源读/写的权限。

```
[root@ controller ~] # rabbitmqctl set_ permissions
openstack ". * " ". * " ". * "
```

②查询 OpenStack 用户所拥有的权限。

```
[root@ controller ~] # rabbitmqctl list_ user_
permissions openstack
```

结果如图 3-5 所示。

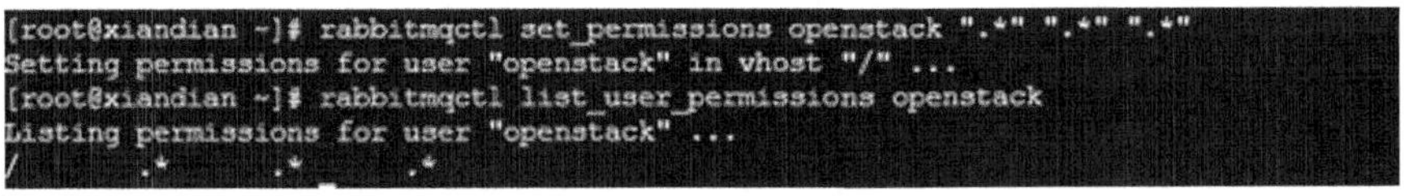

```
[root@xiandian ~]# rabbitmqctl set_permissions openstack ".*" ".*" ".*"
Setting permissions for user "openstack" in vhost "/" ...
[root@xiandian ~]# rabbitmqctl list_user_permissions openstack
Listing permissions for user "openstack" ...
/       .*      .*      .*
```

图 3-5　权限查询

项目小结

OpenStack 遵循这样的设计原则，服务之间通过 RESTAPI 进行通信。服务内容，不同的服务进程之间的通信，则必须通过消息总线。OpenStack 的 RabbitMQ 实现消息总线的功能。本项目介绍 OpenStack 的消息队列的相关概念和 RabbitMQ 的服务流程和运行机制，而且通过任务使读者较好地完成项目目标。

项目扩展

表 3－3 列举本项目 rabbitmqctl 的常用命令及其说明。

表 3－3　常用命令及其说明

常用命令	命令说明
rabbitmqctl stop	停止 RabbitMQ 应用,关闭节点
rabbitmqctl stop_app	停止 RabbitMQ 应用
rabbitmqctl start_app	启动 RabbitMQ 应用
rabbitmqctl status	显示 RabbitMQ 中间件各种信息
rabbitmqctl reset	重置 RabbitMQ 节点
rabbitmqctl add_user username password	添加用户
rabbitmqctl delete_user username	删除用户
rabbitmqctl change_password username newpassword	修改密码
rabbitmqctl list_users	列出所有用户

项目思考

结合本项目的知识点和网上资料，在 OpenStack 的消息服务中常提到的远程过程调用（RPC）和事件通知（EN）的具体过程是如何进行的？

项目四 镜像服务

项目综述

小张经过一系列学习之后，已经初步了解 OpenStack 消息服务的机制并掌握了相关的基本操作。现在，小张要了解 OpenStack 另外一种服务——镜像服务。小张能够了解镜像服务 Glance 在 OpenStack 整体架构的作用、清楚服务框架流程，并熟练掌握镜像的制作方法。为此，小张当前要完成的任务如下：

视频

项目四综述

- 镜像服务基本操作。
- 制作镜像。
- 云平台的镜像上传。

项目目标

【知识目标】

- OpenStack 的 Glance 服务组件的相关概念。
- OpenStack 的 Glance 的服务流程和运行机制。

【技能目标】

- 镜像服务基本操作。
- 制作镜像。
- 云平台的镜像上传。

【职业能力】

根据企业的应用需求，使用制作的镜像和上传镜像到云平台，并做相应的维护。

任务　安装与制作镜像服务

任务要求

小张经过一系列学习之后，已经初步掌握了 OpenStack 消息服务的机制。现在，小张要了解 OpenStack 另外一种服务——Glance。完成本任务学习后，小张能够了解

Glance 镜像服务在 OpenStack 整体架构的作用、清楚服务框架流程，并熟练掌握镜像的制作方法。

核心概念

镜像服务：在 OpenStack 框架中，镜像服务模块 Glance 实现发现、注册获取虚拟机镜像和镜像元数据，镜像数据支持存储多种存储系统，可以是简单文件系统、对象存储系统等。

知识准备

视频

镜像服务的相关知识

1. 概述

Glance（OpenStack Image Service）是一个提供发现、注册和下载镜像的服务。Glance 提供了虚拟机镜像的集中存储。通过 Glance 的 RESTful API，可以查询镜像元数据、下载镜像。虚拟机的镜像可以很方便地存储在各种地方，从简单的文件系统到对象存储系统（如 OpenStack Swift）。

在 Glance 里镜像被当作模板来存储，用于启动新实例。Glance 还可以从正在运行的实例建立快照用于备份虚拟机的状态。

Glance 具体功能如下：

（1）提供 RESTful API 让用户能够查询和获取镜像的元数据和镜像本身。

（2）支持多种方式存储镜像，包括普通的文件系统、Swift、Ceph 等。

（3）对实例执行快照创建新的镜像。

2. Glance 服务架构

Glance 镜像服务架构是典型的 C/S 架构，Glance 架构包括 Glance - Client、Glance 和 Glance Store。Glance 主要包括 REST API、数据库抽象层（Database Abstraction Layer，DAL）、域控制器（Glance Domain Controller）和注册层（Registry Layer），Glance 使用集中数据库（Glance DB）在 Glance 各组件间直接共享数据。所有的镜像文件操作都通过 Glance Store 库完成，Glance Store 库提供了通用接口，对接后端外部不同存储。Glance 架构如图 4 - 1 所示。

（1）客户端（Client）：外部用于同 Glance 服务的交互和操作。

（2）Glance - API：Glance 对外的 REST API 接口。

（3）数据库抽象层（DAL）：Glance 和数据库直接交互的编程接口。

（4）Glance 域控制器：中间件实现 Glance 的认证、通知、策略和数据链接等主要功能。

（5）注册层：可选层，用于管理域控制和数据库 DAL 层之间的安全通信。

（6）Glance DB：存储镜像的元数据，根据需要可以选择不同类型的数据库，目前采用 MySQL。

（7）Glance Store：Glance 对接不同数据存储的抽象层。

（8）后端存储：实际接入的存储系统。可以接入简单文件系统、Swift、Ceph 和 S3 云存储等。当前框架选择存储在本地，目录在控制节点：/var/lib/glance/image。

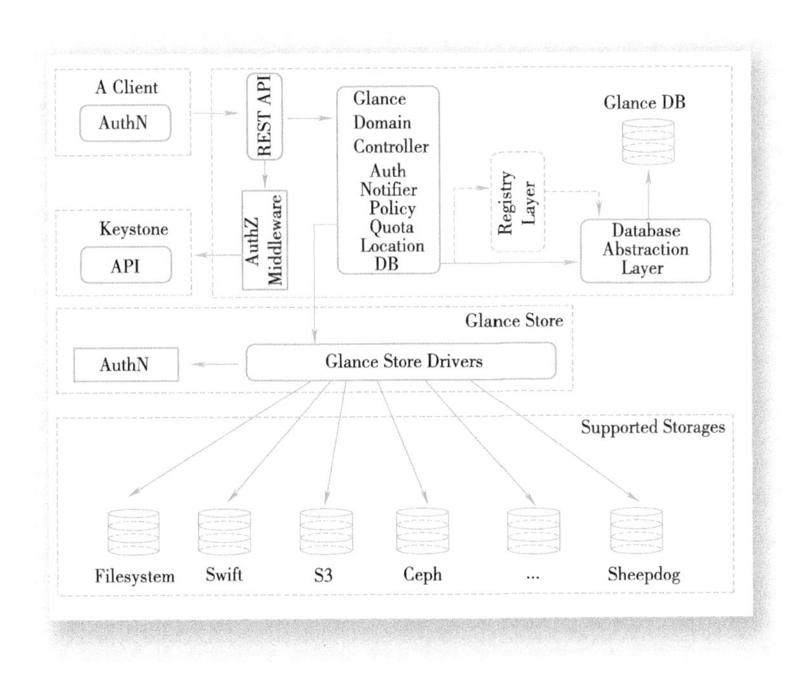

图4-1 Glance架构

Glance服务自带两个配置文件，在使用Glance镜像服务时需要配置Glance-Registry conf两个服务模块。在添加镜像到Glance时，需要指定虚拟镜像的磁盘的磁盘格式（Disk Format）和容器格式（Container Format）。

镜像服务运行两个后台服务进程Demon如下：

（1）Glance-API：对外接受并发布镜像、获取和存储的请求调用。

（2）Glance-Registry：存储、处理和获取镜像元数据，内部服务只有Glance内部使用，不暴露给用户。

（3）Glance-All：是对前两个进程的通用封装，操作方式和结果一样。

3. 镜像文件格式

虚拟机镜像需要指定磁盘格式和容器格式。虚拟设备供应商将不同的格式存储在一个虚拟机磁盘映像中，虚拟机的磁盘镜像的格式基本有如下几种：

（1）raw：非结构化磁盘镜像格式。

（2）qcow2：QEMU模拟器支持的可动态扩展、写时复制的磁盘格式，是KVM虚拟默认使用的磁盘文件格式。

（3）AMI/AKI/ARI：Amazon EC2最初支持的镜像格式。

（4）UEC tarball：Ubuntu Enterprise Cloud tarball是一个gzip压缩后的tar文件。

（5）VHD：Microsoft Virtual Hard Disk Format（微软虚拟磁盘文件）的简称。

（6）VD：VirtualBox使用VDI（Virtual Disk Image）的镜像格式，OpenStack没有提供直接的支持，需要进行格式转换。

（7）VMDK：VMware Virtual Machine Disk Format是虚拟机VMware创建的虚拟机格式。

（8）OVF（Open Virtualization Format，开放虚拟化格式）：OVF文件是一种开源的文件规范，可用于虚拟机文件的打包。

容器格式是可以理解成虚拟机镜像添加元数据后重新打包的格式，有以下几种容器格式：

（1）Bare：指定没有容器和元数据封装在镜像中，如果 Glance 和 OpenStack 的其他服务没有使用容器格式的字符串，为了安全，建议设置 bare。

（2）ovf：ovf 的容器模式。

（3）aki：存储在 Glance 中的是 Amazon 的内核镜像。

（4）ari：存储在 Glance 中的是 Amazon 的 ramdisk 镜像。

（5）ami：存储在 Glance 中的是 Amazon 的 machine 镜像。

（6）ova：存储在 Glance 中的是 OVA 的 tar 归档文件。

4. 镜像状态

OpenStack 的镜像有如下多种状态，这些状态之间可以相互转换：

（1）queued 排队：镜像已注册。image 数据还没有上传到 Glance，image 在创建过程中，大小没有明确设置为 0。

（2）saving 保存中：镜像的原始数据正在被上传到 Glance。

（3）active 有效：镜像在 Glance 上是完全可用的。当镜像已经完全上传完成，或者镜像大小在创建阶段被设置为 0，镜像成功创建，状态有效可用。

（4）killed 错误：上传镜像数据出错，目前不可读取。

（5）deleted 被删除：镜像不可用，将被自动删除。

（6）pending_delete 等着删除：镜像不可用，等待将被自动删除。

由于镜像文件都比较大，镜像从创建到成功上传至 Glance 文件系统中，是通过异步任务的方式一步步完成，如图 4－2所示。

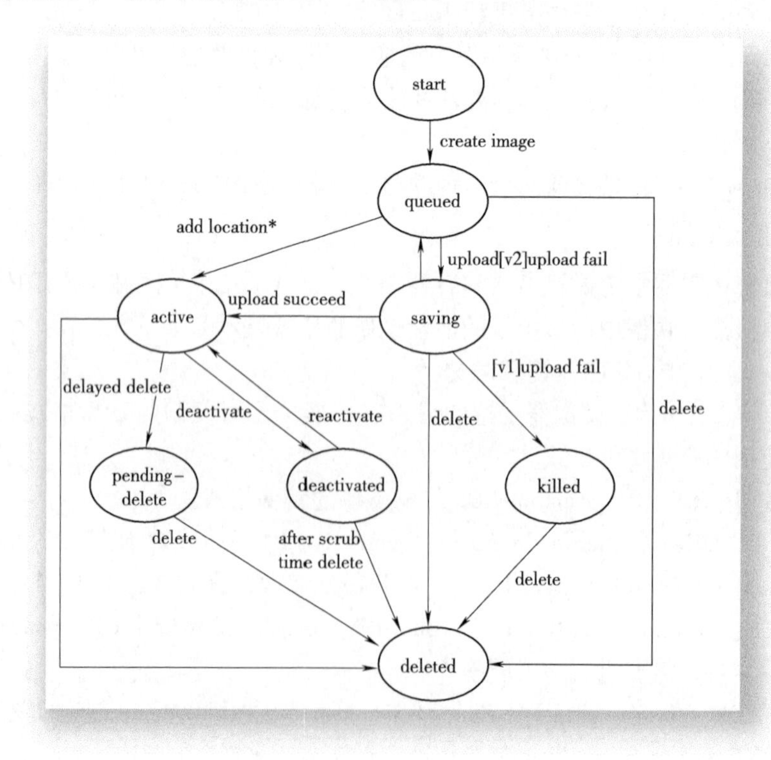

图 4－2　镜像状态图

镜像服务的基本操作

1. 准备实施任务的环境（见表4-1）

表4-1 实施任务的准备环境

实施任务所需软件资源	虚拟机镜像资源信息
虚拟机软件（vm12）	controller
securcrt远程连接软件	compute

2. 分解实施任务的过程（见表4-2）

表4-2 实施任务的简明过程

序号	步骤	详细操作及说明
1	镜像服务基本操作	①命令行方式进行镜像创建、查询、删除和修改镜像； ②创建镜像； ③更改镜像
2	制作镜像	①挂载系统的iso文件； ②安装虚拟化工具软件包； ③启动libvirtd； ④使用KVM创建CentOS7的虚拟机
3	云平台的镜像上传	①Dashboard界面上传镜像； ②命令行CLI上传镜像

3. 镜像服务基本操作

(1) 命令行方式进行镜像创建、查询、删除和修改镜像。

①查询Glance版本。

检测Glance服务列表。

```
[root@ controller ~] #source /etc/keystone/admin-openrc.sh
[root@ controller ~] #openstack-service list |grep glance
```

结果如图4-3所示。

```
[root@xiandian ~]# openstack-service list |grep glance
openstack-glance-api
openstack-glance-registry
```

图4-3 服务列表

②检测Glance服务是否启动。

```
[root@ controller ~] #openstack-service status |grep glance
```

结果如图4-4所示。

```
[root@xiandian ~]# openstack-service status |grep glance
MainPID=935 Id=openstack-glance-api.service ActiveState=active
MainPID=929 Id=openstack-glance-registry.service ActiveState=active
```

图 4-4　服务状态

③查询 glance-control 版本。

```
[root@ controller ~] # glance-control --version
```

结果如图 4-5 所示。

```
[root@xiandian ~]# glance-control --version
12.0.0
```

图 4-5　版本信息

（2）创建镜像。

①使用终端软件上传 CirrOS 镜像到 controller 的/tmp/images 目录中并查看。

```
[root@ controller ~] # mkdir /tmp/images
[root@ controller ~] # cd /tmp/images/
```

结果如图 4-6 所示。

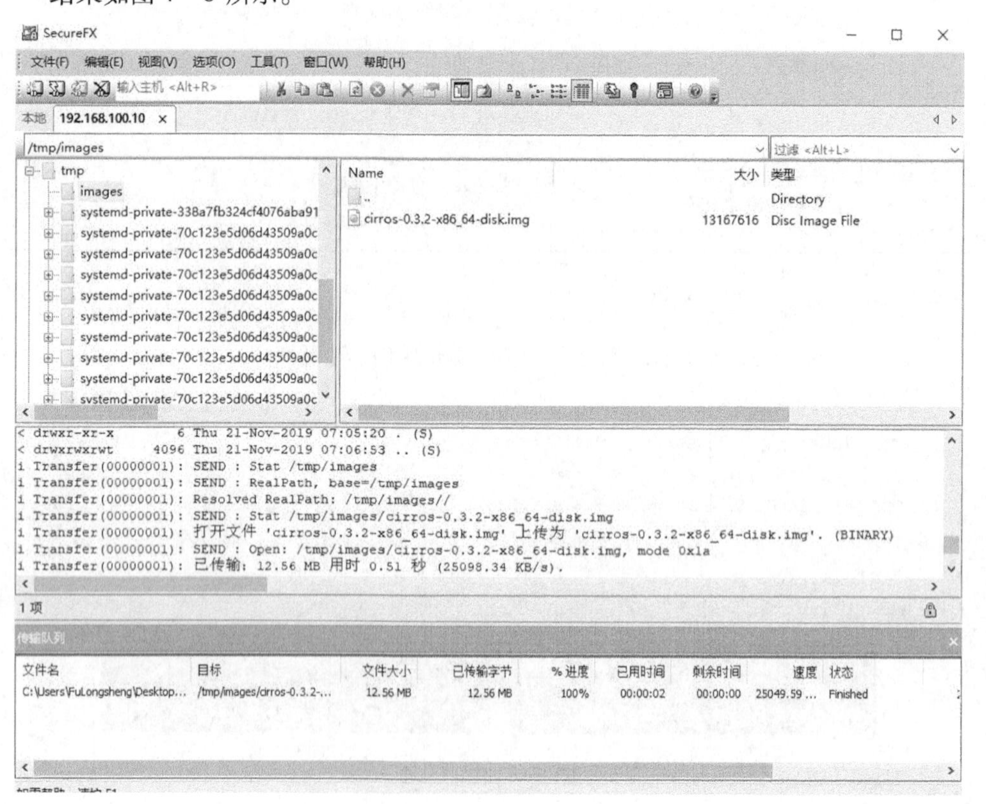

图 4-6　拉取情况

②查看镜像文件信息。

```
[root@ controller images] # file cirros-0.3.2-x86_64-disk.img
```

结果如图 4 –7 所示。

```
[root@xiandian images]# file cirros-0.3.2-x86_64-disk.img
cirros-0.3.2-x86_64-disk.img: QEMU QCOW Image (v2), 41126400 bytes
```

图 4 –7　文件信息

③使用命令行创建镜像。

```
[root@ controller images] #source /etc/keystone/admin-openrc. sh
[root@ controller images] #glance image-create --name "cirros-0.3.2-x86_64" --disk-format qcow2 --container-format bare --progress < cirros-0.3.2-x86_64-disk. img
```

结果如图 4 –8 所示。

```
[root@controller images]# source /etc/keystone/admin-openrc.sh
[root@controller images]# glance image-create --name "cirros-0.3.2-x86_64" --disk-format qcow2 --container-format bare --progress < cirros-0.3.2-x86_64-disk.img
[=============================>] 100%
+------------------+--------------------------------------+
| Property         | Value                                |
+------------------+--------------------------------------+
| checksum         | 64d7c1cd2b6f60c92c14662941cb7913     |
| container_format | bare                                 |
| created_at       | 2019-12-13T03:10:09Z                 |
| disk_format      | qcow2                                |
| id               | c5db9326-8d67-43e8-82cd-6db9fc39d36b |
| min_disk         | 0                                    |
| min_ram          | 0                                    |
| name             | cirros-0.3.2-x86_64                  |
| owner            | d2b88800df664a8cb196ec75034c9e45     |
| protected        | False                                |
| size             | 13167616                             |
| status           | active                               |
| tags             | []                                   |
| updated_at       | 2019-12-13T03:10:09Z                 |
| virtual_size     | None                                 |
| visibility       | private                              |
```

图 4 –8　创建镜像

④查询镜像列表。

```
[root@ controller images] #glance image-list
```

结果如图 4 –9 所示。

```
[root@xiandian images]# glance image-list
+--------------------------------------+---------------------+
| ID                                   | Name                |
+--------------------------------------+---------------------+
| f837fedd-1d38-4f16-8708-8e9cc716de78 | cirros-0.3.2-x86_64 |
+--------------------------------------+---------------------+
```

图 4 –9　镜像列表

（3）更改镜像。

可以使用 Glance image – update 更新镜像信息，可以使用 Glance image – delete 删除镜像信息。如果需要改变镜像启动硬盘最低要求值（min – disk）时，min – disk 默认单位为 G。

①获取镜像详细信息。镜像的 ID 通过镜像列表查询得出，每个镜像的 ID 都不同。

```
[root@ controller images] #glance image-list
[root@ controller images] #glance image-show f837fedd-1d38-4f16-8708-8e9cc716de78
```

结果如图 4 –10 所示。

```
[root@xiandian images]# glance image-list
+--------------------------------------+---------------------+
| ID                                   | Name                |
+--------------------------------------+---------------------+
| f837fedd-1d38-4f16-8708-8e9cc716de78 | cirros-0.3.2-x86_64 |
+--------------------------------------+---------------------+
[root@xiandian images]# glance image-show f837fedd-1d38-4f16-8708-8e9cc716de78
+------------------+--------------------------------------+
| Property         | Value                                |
+------------------+--------------------------------------+
| checksum         | 64d7c1cd2b6f60c92c14662941cb7913     |
| container_format | bare                                 |
| created_at       | 2019-06-03T06:47:11Z                 |
| disk_format      | qcow2                                |
| id               | f837fedd-1d38-4f16-8708-8e9cc716de78 |
| min_disk         | 0                                    |
| min_ram          | 0                                    |
| name             | cirros-0.3.2-x86_64                  |
| owner            | 0ab2dbde4f754b699e22461426cd0774     |
| protected        | False                                |
| size             | 13167616                             |
| status           | active                               |
| tags             | []                                   |
| updated_at       | 2019-06-03T06:47:11Z                 |
| virtual_size     | None                                 |
| visibility       | private                              |
+------------------+--------------------------------------+
```

图 4-10 详细信息

②修改镜像启动硬盘所需大小。

```
[root@ controller images] # glance image - update -- min - disk =1 f837fedd -1d38 -4f16 -8708 -8e9cc716de78
```

结果如图 4-11 所示。

```
[root@xiandian images]# glance image-update --min-disk=1 f837fedd-1d38-4f16-8708-8e9cc716de78
+------------------+--------------------------------------+
| Property         | Value                                |
+------------------+--------------------------------------+
| checksum         | 64d7c1cd2b6f60c92c14662941cb7913     |
| container_format | bare                                 |
| created_at       | 2019-06-03T06:47:11Z                 |
| disk_format      | qcow2                                |
| id               | f837fedd-1d38-4f16-8708-8e9cc716de78 |
| min_disk         | 1                                    |
| min_ram          | 0                                    |
| name             | cirros-0.3.2-x86_64                  |
| owner            | 0ab2dbde4f754b699e22461426cd0774     |
| protected        | False                                |
| size             | 13167616                             |
| status           | active                               |
| tags             | []                                   |
| updated_at       | 2019-06-03T06:57:13Z                 |
| virtual_size     | None                                 |
| visibility       | private                              |
+------------------+--------------------------------------+
```

图 4-11 更新镜像

③删除镜像。

```
[root@ controller images] # glance image - delete f837fedd -1d38 -4f16 -8708 -8e9cc716de78
[root@ controller images] # glance image - list
```

结果如图 4－12 所示。

```
[root@xiandian images]# glance image-delete f837fedd-1d38-4f16-8708-8e9cc716de78
[root@xiandian images]# glance image-list
+----+------+
| ID | Name |
+----+------+
+----+------+
```

图 4－12　镜像列表

4. 制作 CentOS 7.2 镜像

视频

制作CentOS7.2镜像

(1)挂载 CentOS 7.2 的 iso 文件,如果在/opt 目录下有相应的目录就不用挂载,如图 4－13 所示。

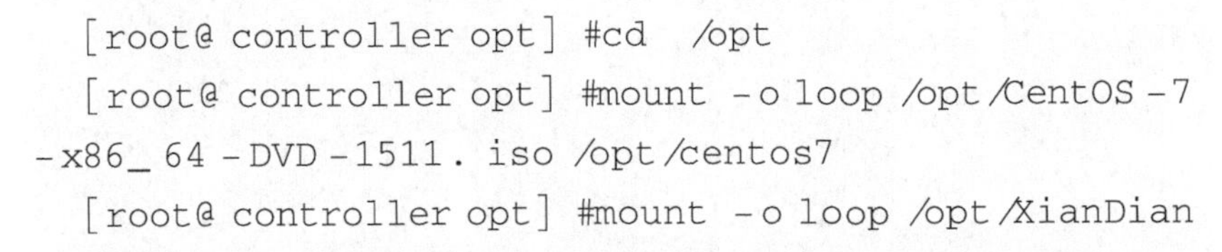

```
[root@ controller opt] #cd  /opt
[root@ controller opt] #mount -o loop /opt/CentOS-7-x86_64-DVD-1511.iso /opt/centos7
[root@ controller opt] #mount -o loop /opt/XianDian-IaaS-v2.2.iso/opt/iaas
```

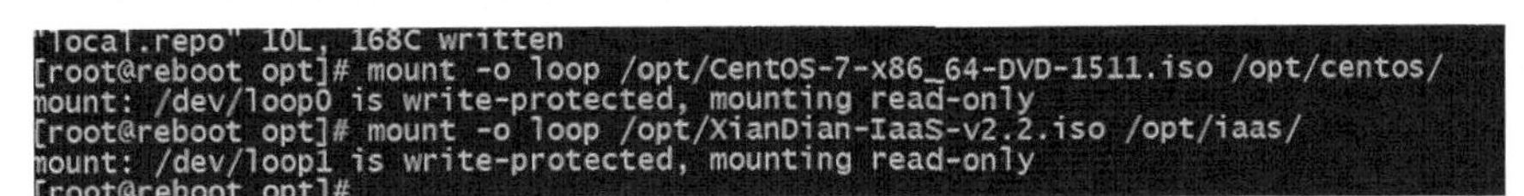

```
"local.repo" 10L, 168C written
[root@reboot opt]# mount -o loop /opt/CentOS-7-x86_64-DVD-1511.iso /opt/centos/
mount: /dev/loop0 is write-protected, mounting read-only
[root@reboot opt]# mount -o loop /opt/XianDian-IaaS-v2.2.iso /opt/iaas/
mount: /dev/loop1 is write-protected, mounting read-only
[root@reboot opt]#
```

图 4－13　挂载 CentOS 7.2 的 iso 文件

(2) 安装虚拟化工具软件包 qemu－kvm 和 libvirt，如图 4－14 所示。

```
[root@ controller opt] #yum install -y qemu-kvm libvirt
###qemu-kvm 用来创建虚拟机，libvirt 用来管理虚拟机。
```

```
  Verifying  : 2:libogg-1.3.0-7.el7.x86_64                                17/28
  Verifying  : libICE-1.0.9-2.el7.x86_64                                  18/28
  Verifying  : ipxe-roms-qemu-20160127-1.git6366fa7a.el7.noarch           19/28
  Verifying  : seavgabios-bin-1.7.5-11.el7.noarch                         20/28
  Verifying  : 1:libvorbis-1.3.3-8.el7.x86_64                             21/28
  Verifying  : libXi-1.7.4-2.el7.x86_64                                   22/28
  Verifying  : seabios-bin-1.7.5-11.el7.noarch                            23/28
  Verifying  : libXau-1.0.8-2.1.el7.x86_64                                24/28
  Verifying  : libSM-1.2.2-2.el7.x86_64                                   25/28
  Verifying  : celt051-0.5.1.3-8.el7.x86_64                               26/28
  Verifying  : libsndfile-1.0.25-10.el7.x86_64                            27/28
  Verifying  : usbredir-0.6-7.el7.x86_64                                  28/28

Installed:
  qemu-kvm-ev.x86_64 10:2.3.0-31.el7.16.1

Dependency Installed:
  celt051.x86_64 0:0.5.1.3-8.el7              flac-libs.x86_64 0:1.3.0-5.el7_1
  gsm.x86_64 0:1.0.13-11.el7                  ipxe-roms-qemu.noarch 0:20160127-1.git6366fa7a.el7
  libICE.x86_64 0:1.0.9-2.el7                 libSM.x86_64 0:1.2.2-2.el7
  libX11.x86_64 0:1.6.3-2.el7                 libX11-common.noarch 0:1.6.3-2.el7
  libXau.x86_64 0:1.0.8-2.1.el7               libXext.x86_64 0:1.3.3-3.el7
  libXi.x86_64 0:1.7.4-2.el7                  libXtst.x86_64 0:1.2.2-2.1.el7
  libasyncns.x86_64 0:0.8-7.el7               libogg.x86_64 2:1.3.0-7.el7
  libpng.x86_64 2:1.5.13-7.el7_2              libsndfile.x86_64 0:1.0.25-10.el7
  libusbx.x86_64 0:1.0.15-4.el7               libvorbis.x86_64 1:1.3.3-8.el7
  libxcb.x86_64 0:1.11-4.el7                  pixman.x86_64 0:0.32.6-3.el7
  pulseaudio-libs.x86_64 0:6.0-7.el7          qemu-kvm-common-ev.x86_64 10:2.3.0-31.el7.16.1
  seabios-bin.noarch 0:1.7.5-11.el7           seavgabios-bin.noarch 0:1.7.5-11.el7
  sgabios-bin.noarch 1:0.20110622svn-4.el7    spice-server.x86_64 0:0.12.4-15.el7_2.2
  usbredir.x86_64 0:0.6-7.el7

Complete!
[root@controller ~]#
```

图 4－14　安装虚拟化工具软件包 1

(3) 安装虚拟化工具软件包 virt－install，用来创建虚拟机，如图 4－15 所示。

```
[root@ controller opt] #yum install -y virt-install
```

```
---> Package gnutls.x86_64 0:3.3.8-12.el7_1.1 will be updated
---> Package gnutls.x86_64 0:3.3.8-14.el7_2 will be an update
---> Package gnutls-dane.x86_64 0:3.3.8-14.el7_2 will be installed
---> Package unbound-libs.x86_64 0:1.4.20-26.el7 will be installed
--> Processing Dependency: libldns.so.1()(64bit) for package: unbound-libs-1.4.20-26.el7.x86_64
--> Running transaction check
---> Package ldns.x86_64 0:1.6.16-7.el7 will be installed
--> Finished Dependency Resolution

Dependencies Resolved

================================================================================================
 Package                     Arch           Version                    Repository         Size
================================================================================================
Installing:
 virt-install                noarch         1.2.1-8.el7                centos             85 k
Installing for dependencies:
 cyrus-sasl                  x86_64         2.1.26-20.el7_2            iaas               88 k
 cyrus-sasl-md5              x86_64         2.1.26-20.el7_2            iaas               56 k
 gnutls-dane                 x86_64         3.3.8-14.el7_2             iaas               32 k
 gnutls-utils                x86_64         3.3.8-14.el7_2             iaas              228 k
 ldns                        x86_64         1.6.16-7.el7               centos            473 k
 libosinfo                   x86_64         0.2.12-3.el7               centos            256 k
 libvirt-client              x86_64         1.2.17-13.el7_2.5          iaas              4.3 M
 libvirt-python              x86_64         1.2.17-2.el7               centos            309 k
 nmap-ncat                   x86_64         2:6.40-7.el7               centos            201 k
 python-ipaddr               noarch         2.1.9-5.el7                centos             36 k
 unbound-libs                x86_64         1.4.20-26.el7              centos            296 k
 virt-manager-common         noarch         1.2.1-8.el7                centos            1.0 M
 yajl                        x86_64         2.0.4-4.el7                centos             39 k
Updating for dependencies:
 cyrus-sasl-lib              x86_64         2.1.26-20.el7_2            iaas              155 k
 gnutls                      x86_64         3.3.8-14.el7_2             iaas              662 k
```

图 4－15 安装虚拟化工具软件包 2

（4）启动 libvirtd。

启动 libvirtd，并将它设为开机启动，启动后使用 ifconfig 查看，发现多出来一块 virbr 0 的网卡，说明 libvirtd 启动成功，如果默认没有 ifconfig 命令，使用 yum install -y net-tools 安装，如果能看见 virbr 0，就表示成功。结果如图 4－16 所示。

```
[root@ controller bin] # cd  /usr/local/bin
[root@ controller bin] # systemctl start libvirtd && systemctl enable libvirtd
[root@ controller bin] # ifconfig
```

```
Complete!
[root@controller bin]# systemctl start libvirtd && systemctl enable libvirtd
[root@controller bin]# ifconfig
eno16777736: flags=4163<UP,BROADCAST,RUNNING,MULTICAST>  mtu 1500
        inet 192.168.100.10  netmask 255.255.255.0  broadcast 192.168.100.255
        inet6 fe80::20c:29ff:fe99:ecf9  prefixlen 64  scopeid 0x20<link>
        ether 00:0c:29:99:ec:f9  txqueuelen 1000  (Ethernet)
        RX packets 5239194  bytes 7488151906 (6.9 GiB)
        RX errors 0  dropped 0  overruns 0  frame 0
        TX packets 531671  bytes 178886787 (170.5 MiB)
        TX errors 0  dropped 0 overruns 0  carrier 0  collisions 0

eno33554960: flags=4163<UP,BROADCAST,RUNNING,MULTICAST>  mtu 1500
        inet6 fe80::20c:29ff:fe99:ec03  prefixlen 64  scopeid 0x20<link>
        ether 00:0c:29:99:ec:03  txqueuelen 1000  (Ethernet)
        RX packets 111  bytes 10086 (9.8 KiB)
        RX errors 0  dropped 0  overruns 0  frame 0
        TX packets 32  bytes 2460 (2.4 KiB)
        TX errors 0  dropped 0 overruns 0  carrier 0  collisions 0

lo: flags=73<UP,LOOPBACK,RUNNING>  mtu 65536
        inet 127.0.0.1  netmask 255.0.0.0
        inet6 ::1  prefixlen 128  scopeid 0x10<host>
        loop  txqueuelen 0  (Local Loopback)
        RX packets 65375  bytes 280626822 (267.6 MiB)
        RX errors 0  dropped 0  overruns 0  frame 0
        TX packets 65375  bytes 280626822 (267.6 MiB)
        TX errors 0  dropped 0 overruns 0  carrier 0  collisions 0

virbr0: flags=4099<UP,BROADCAST,MULTICAST>  mtu 1500
        inet 192.168.122.1  netmask 255.255.255.0  broadcast 192.168.122.255
        ether 52:54:00:68:92:fa  txqueuelen 0  (Ethernet)
        RX packets 0  bytes 0 (0.0 B)
        RX errors 0  dropped 0  overruns 0  frame 0
        TX packets 0  bytes 0 (0.0 B)
        TX errors 0  dropped 0 overruns 0  carrier 0  collisions 0

[root@controller bin]#
```

图 4－16 启用 libvirtd

（5）使用 KVM 创建 CentOS 7 的虚拟机。

①使用 qemu 命令创建一个 10 GB 的硬盘的虚拟机（最小 10 GB，可以更多），虚拟机的名称为：CentOS－7－x86_64. raw，如图 4－17 所示。

```
[root@ controller ~] #qemu-img create -f raw /opt/CentOS-7-x86_64.raw 10G
[root@ controller ~] #ll -h /opt
```

```
[root@controller ~]# qemu-img create -f raw /opt/CentOS-7-x86_64.raw 10G
Formatting '/opt/CentOS-7-x86_64.raw', fmt=raw size=10737418240
[root@controller ~]# qemu-img create -f raw /opt/CentOS-7-x86_64.raw 10G
Formatting '/opt/CentOS-7-x86_64.raw', fmt=raw size=10737418240
[root@controller ~]# ll -h /opt
total 6.7G
dr-xr-xr-x. 8 root root 2.0K Dec 10  2015 centos
-rw-r--r--. 1 root root 4.1G May 18  2018 CentOS-7-x86_64-DVD-1511.iso
-rw-r--r--. 1 root root  10G Apr 26 22:50 CentOS-7-x86_64.raw
drwxr-xr-x. 4 root root 2.0K Nov  6  2017 iaas
-rw-r--r--. 1 root root 1022 Apr 26 20:40 server.yml
-rw-r--r--. 1 root root 2.7G Nov  6  2017 XianDian-IaaS-v2.2.iso
[root@controller ~]#
```

图 4－17　qemu 命令创建硬盘

②使用 virt－install 创建名称为 CentOS－7－x86_ 64 的虚拟机，在创建之前，先上传一个 CentOS 7 的 ISO 镜像，如图 4－18 所示。

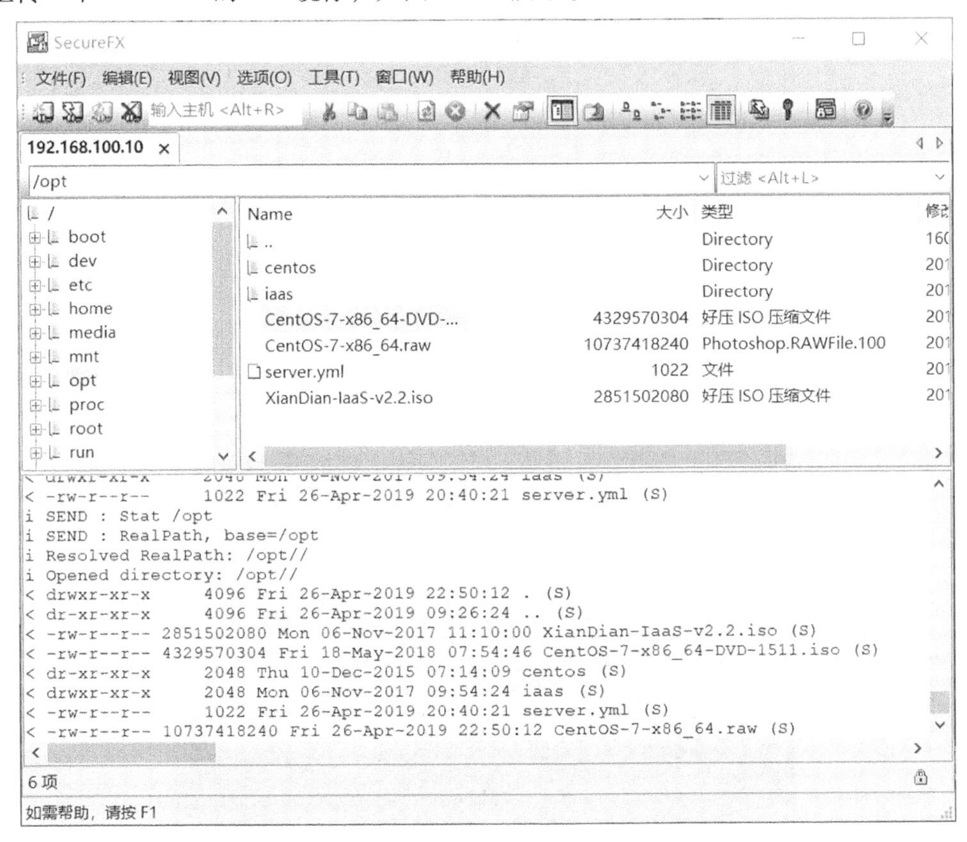

图 4－18　使用 virt－install 创建名称为 CentOS－7－x86_ 64 的虚拟机

③创建虚拟机，如图 4－19 所示。

```
[root@ controller ~] # virt - install  -- virt - type kvm  --
name CentOS - 7  - x86_ 64  -- ram 1024  -- cdrom = /opt /CentOS - 7 -
x86_ 64 - DVD - 1511. iso  -- disk path = /opt /CentOS - 7 - x86_ 64.
raw  -- network network = default  -- graphics vnc, listen = 0. 0.
0. 0  -- noautoconsole
```

```
Last login: Thu Aug 29 17:50:19 2019 from 192.168.100.7
[root@controller ~]# virt-install --virt-type kvm --name CentOS-7-x86_64 --ram 1024 --cdrom=/opt
/CentOS-7-x86_64-DVD-1511.iso --disk path=/opt/CentOS-7-x86_64.raw --network network=default --g
raphics vnc,listen=0.0.0.0 --noautoconsole

Starting install...
Creating domain...                                                     |    0 B  00:00:07
Domain installation still in progress. You can reconnect to
the console to complete the installation process.
[root@controller ~]#
```

图 4 – 19　创建虚拟机

④使用 TightVNC 工具，连接主机 IP 192. 168. 100. 10，设置安装操作系统的网卡名称为 eth0，如图 4 – 20 和图 4 – 21 所示。

New TightVNC Connection

Connection

Remote Host: 192.168.100.10　　Connect

Enter a name or an IP address. To specify a port number, append it after two colons (for example, mypc::5902).　　Options...

Reverse Connections

Listening mode allows people to attach your viewer to their desktops. Viewer will wait for incoming connections.　　Listening mode

TightVNC Viewer

TightVNC is cross-platform remote control software.

Its source code is available to everyone, either freely (GNU GPL license) or commercially (with no GPL restrictions).

Version info...　　Licensing　　Configure...

图 4 – 20　VNC 连接

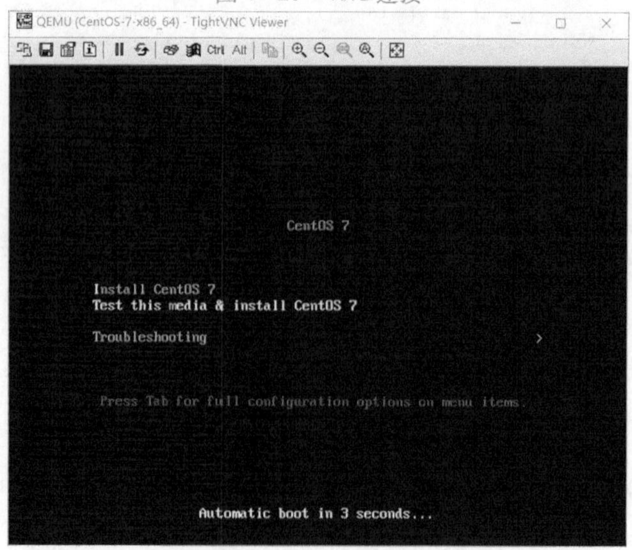

图 4 – 21　CentOS 7 启用

安装步骤与安装操作系统的方法一样，安装完成后，可以使用 virsh list --all 显示 KVM 上所有的虚拟机，如图 4-22 所示。

```
[root@ controller ~] #virsh list --all
```

可以看到虚拟机的名称和状态。

```
[root@controller ~]# virsh list --all
 Id    Name                           State
----------------------------------------------------
 -     CentOS-7-x86_64                shut off

[root@controller ~]#
```

图 4-22　显示 KVM 上所有的虚拟机

⑤管理 KVM。

a. 使用 virsh 启动 KVM 中的虚拟机。

```
[root@ controller ~] #virsh start CentOS-7-x86_64
```

启动后，使用 VNC 连接工具重新登录，如图 4-23 所示。

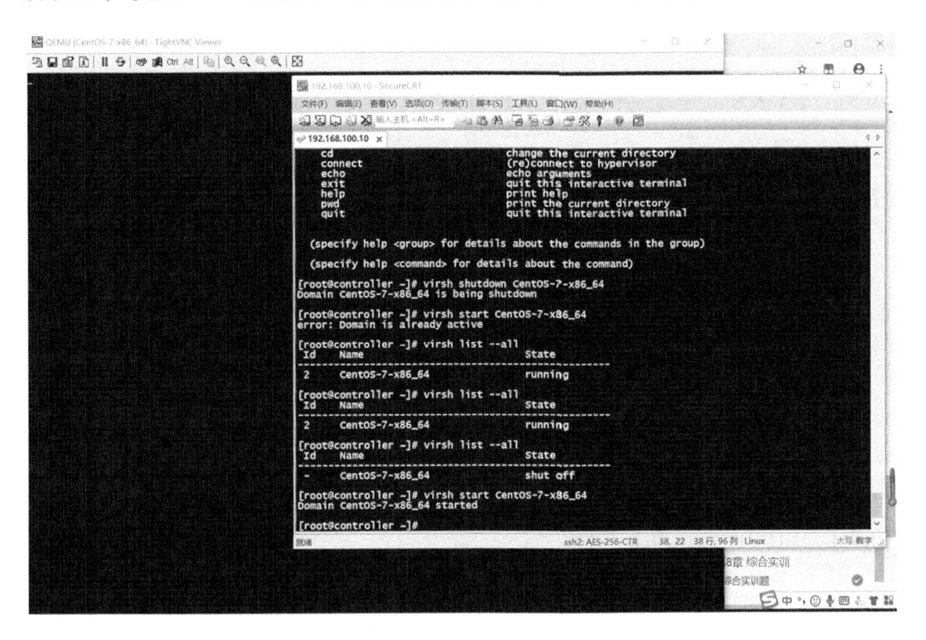

图 4-23　VNC 连接工具重新登录虚拟机

b. 配置网卡 IP 地址、重启网卡，如图 4-24 和图 4-25 所示。

```
[root@ localhost ~] #vi /etc/sysconfig/network-scripts/ifcfg-eth0
[root@ localhost ~] #service network restart
```

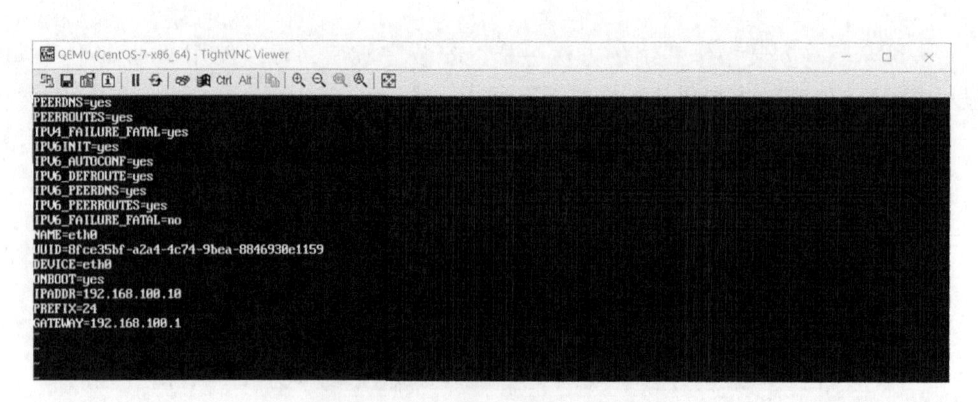

图 4－24　配置网卡 IP 地址

```
"/etc/sysconfig/network-scripts/ifcfg-eth0" 19L, 331C written
[root@localhost ~]# service  network restart
Restarting network (via systemctl):                        [  OK  ]
[root@localhost ~]#
```

图 4－25　重启网卡

c. 关闭虚拟机，如图 4－26 所示。

```
[root@ localhost ~] # Poweroff
```

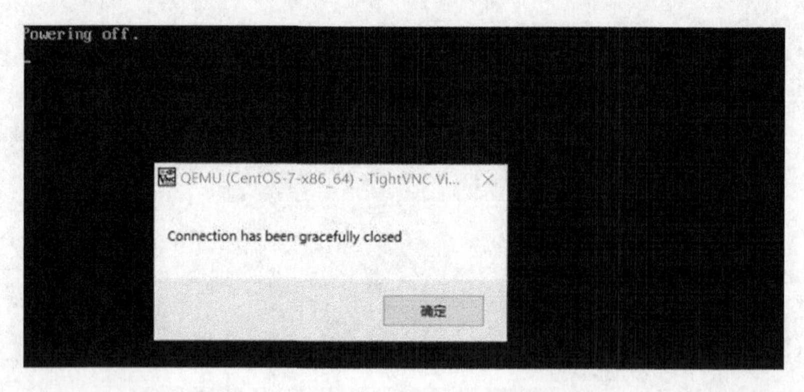

图 4－26　关闭虚拟机

d. 镜像格式转换，如图 4－27 所示。

```
[root@ controller ~] # qemu-img convert -c -O qcow2 /opt/CentOS-7-x86_64.raw /opt/CentOS-7-x86_64.qcow2
```

```
[root@controller ~]# qemu-img convert -c -O qcow2 /opt/CentOS-7-x86_64.raw /opt/CentOS-7-x86_64.qcow2
[root@controller ~]#
```

图 4－27　镜像格式转换

5. 云平台的镜像上传

（1）Dashboard 界面上传镜像，如图 4－28 和图 4－29 所示。

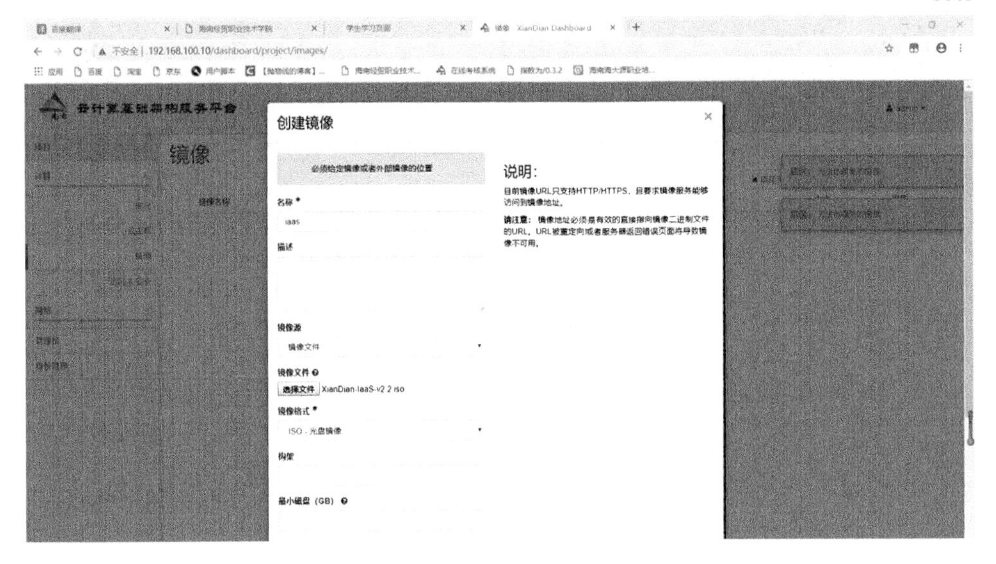

图 4－28　Dashboard 界面上传镜像 1

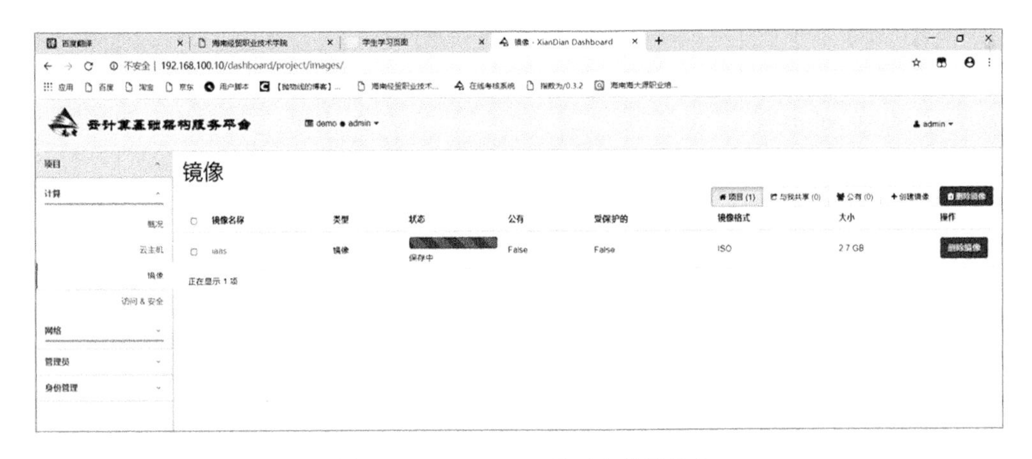

图 4－29　Dashboard 界面上传镜像 2

（2）命令行 CLI 上传镜像，如图 4－30 所示。

```
[root@controller ~]# source /etc/keystone/admin-openrc.sh
[root@controller ~]# glance images-create --name "Centos7" --disk-format=qcow2 --container-format bare --progress < /opt/Centos-7-x86_64.qcow2
```

图 4-30 命令行 CLI 上传镜像

项目小结

在 OpenStack 中，镜像服务 Glance 实现存储镜像数据，给虚拟机使用。本项目讲解了 Glance 服务组件的相关概念、服务流程和运行机制，并通过镜像服务的基本操作、制作镜像和上传镜像到云平台的任务，使读者较好地完成项目目标。

项目扩展

表 4-3 列举本项目 Glance 的常用命令及其说明。

表 4-3 常用命令及其说明

常用命令	命令说明
glance image-list	列出全部镜像
glance image-show	查看镜像详细信息
glance image-create	创建镜像
glance image-delete	删除镜像
glance image-update	更新镜像
glance image-download	下载镜像

项目思考

Glance 为 OpenStack 提供虚拟机的镜像服务。Glance 本身却并不负责实际的存储，只是完成一些镜像工作，结合本项目的知识点，请思考实际的存储工作是由哪些组件完成的。

项目五 网络服务

项目综述

经过一系列学习之后，小张已经初步掌握了认证服务、消息服务及镜像服务，要继续学习网络服务，在 OpenStack 中配置网络是一个令人困惑的经历。因为它提供了一个相当数量的功能和灵活的产品插件来支持虚拟网络。小张需要了解网络服务的相关理论知识，熟悉 Neutron 网络管理命令。然后按照企业的应用需求，创建企业内部网络，满足企业正常的办公需求。小张为此要完成的任务如下：

项目五综述

- 网络服务基本操作。
- 网络的隔离操作。
- 创建符合企业需求的 Neutron 网络。

项目目标

【知识目标】
- OpenStack 的 Neutron 服务组件的相关概念。
- OpenStack 的 Neutron 的服务流程和运行机制。

【技能目标】
- 网络服务基本操作。
- 创建符合企业需求的 Neutron 网络。

【职业能力】

根据企业的应用需求，创建企业内部网络，满足企业正常的办公需求。

任务　安装与操作网络服务

任务要求

在 OpenStack 中配置网络是一个令人困惑的经历。它提供了一个相当数量的功能和灵活的产品插件来支持虚拟网络。接下来，小张需要了解服务的相关理论知识，熟悉 Neutron 网络管理命令。然后按照企业的需求，创建企业内部网络，满足企业正常的办公需求。

公司要求如下：创建项目研发部内部使用网络，名称为 RD－Net，子网名为 RD－Subnet 网段为 172.24.3.0/24，网关为 172.24.3.1。创建业务部内部使用网络，名称为 BS－Net，子网名为 BS－Subnet，网段为 172.24.4.0/24，网关为 172.24.4.1。创建 IT 工程部内部使用网络，名称为 IT－Net，子网名为 IT－Subnet，网段为 172.24.5.0/24，网关为 172.24.5.1。创建外来使用网络名称为 Guest－Net，子网名为 Guest－Subnet，网段为 172.24.6.0/24，网关为 172.24.6.1。

核心概念

网络服务：OpenStack 网络服务 Neutron 为 OpenStack 环境中的虚拟网络基础架构和物理网络基础架构接入层管理所有的网络现状。Neutron 网络提供了以下对象抽象：网络、子网和路由器。每一个都有模仿其物理副本的功能：网络包含子网，路由器在不同的子网和网络之间进行路由通信。

视频

网络服务的相关知识

知识准备

1. 网络服务 Neutron 概述

OpenStack 发展至今，已经有几十个正式项目，Neutron 属于其中一个核心项目，如图 5－1 所示。

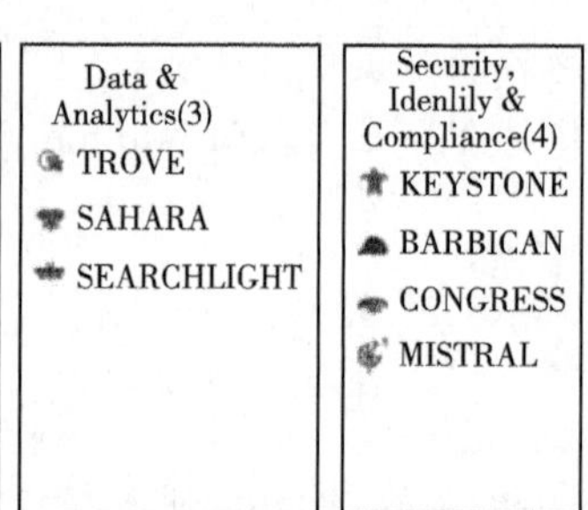

图 5－1　OpenStack 的正式项目

Neutron 在 OpenStack 的主要服务中所处的位置，如图 5－2 所示。

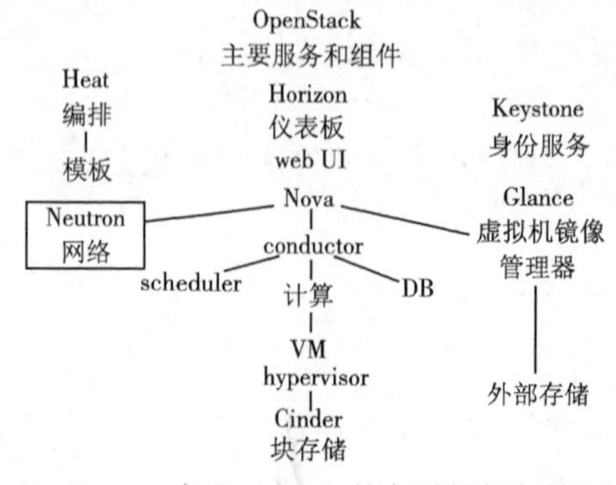

图 5－2　Neutron 在 OpenStack 的主要服务中所处的位置

当前，Neutron 已经成为 OpenStack 三大核心（存储、计算、网络）之一，对外提供NaaS（Network as a Service）服务。但是最初，Neutron只是Nova项目中的一个模块而已，到Folsom版本才正式从中剥离出来，成为一个正式并核心的项目，如表5－1所示。

表5－1 Neutron的发展历史

版本号	发行日期	Neutron历史
Austin	2010.10.21	作为Nova中的一个模块nova－network存在
Essex	2012.04.05	网络功能数据模型开始从Nova中剥离，为独立项目做准备
Folsom	2012.09.27	正式从Nova中剥离，成立新的独立项目Quantum，并且是核心项目
Havana	2013.10.17	项目名称从Quantum改名Neutron

由此，Neutron的发展简史可以概括为三个阶段：Nova－Network、Quantum、Neutron。

Nova－Network阶段，其支持的主要功能有：

（1）IP地址分配：包含为虚拟主机分配私有（固定）和浮动IP地址。

（2）网络管理：仅支持三种网络，扁平网络、带DHCP功能的扁平网络、VLAN网络。

（3）安全控制：主要通过ebtables和iptables来实现。

可以看到，Nova－Network所支持的功能还比较简单。到了Quantum阶段，其支持的主要功能有：

（1）支持多租户隔离，并提供面向租户的API。

（2）插件式结构支持多种网络后端技术，包括Open vSwitch、Cisco、Linux Bridge、Nicira NVP、Ryu、NEC等。

（3）支持位于不同的两层网络的IP地址重叠。

（4）支持基本的三层转发和多路由器。

（5）支持隧道技术（Tunneling）。

（6）支持三层代理和DHCP代理的多节点部署，增强了扩展性和可靠性。

（7）提供负载均衡API（试用版本）。

Quantum阶段所支持的功能已经初具规模。有了Quantum打下的良好基础，进入第三阶段以后，Neutron所支持的功能和应用场景得到了更大的发展。

2. Neutron网络结构

Neutron网络结构如图5－3所示。管理员创建和管理Neutron外部网络，是租户虚拟机与互联网信息交互的桥梁。更具体地说，外部网络会分出一个子网，它是一组在互联网上可寻址的IP地址。一般情况下，外部网络只有一个（Neutron是支持多个外部网络的），且由管理员创建。租户虚拟机创建和管理租户网络，每个网络可以根据需要划分成多个子网。诸多子网通过路由器与Neutron外部网络（图5－3中的子网A）连接。路由器的gateway网关端连接外部网络的子网，interfaces接口端有多个，连接租户网络的子网。路由器及interface接口端连接的网络都是由租户根据需要自助创建，管理者只创建和管理Neutron外部网络部分。

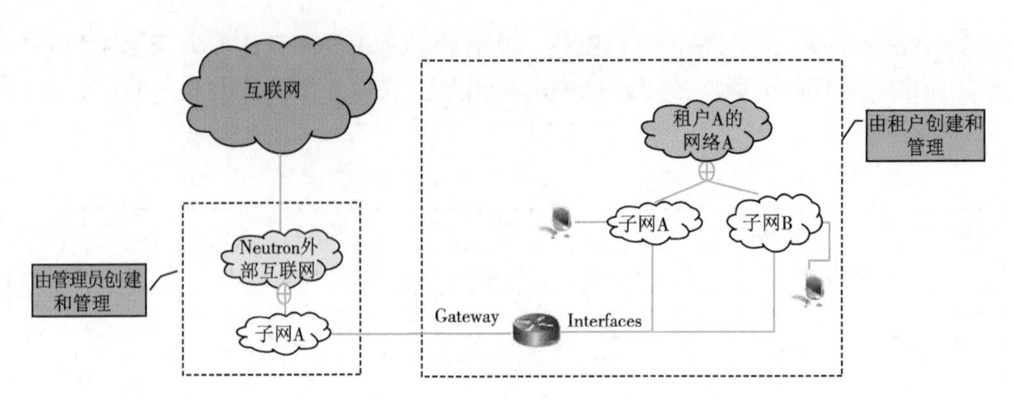

图 5－3　Neutron 网络结构

创建一个 Neutron 网络的过程总结如下：

（1）管理员拿到一组可以在互联网上寻址的 IP 地址，并且创建一个外部网络和子网。

（2）租户创建一个网络和子网。

（3）租户创建一个路由器并且连接租户子网和外部网络。

（4）租户创建虚拟机。

3. Neutron 网络类型

整体网络分为：内部网络（管理网络、数据网络）、外部网络（外部网络、API 网络），如图 5－4 所示。

（1）管理网络（Management Network）：用于 OpenStack 各组件之间的内部通信。

（2）数据网络（Data Network）：用于云部署中虚拟数据之间的通信。

（3）外部网络（Extemal Network）：公共网络，外部或 Internet 可以访问的网络。

（4）API 网络（API Network）：暴露所有的 OpenStack API，包括 OpenStack 网络 API 给租户们。

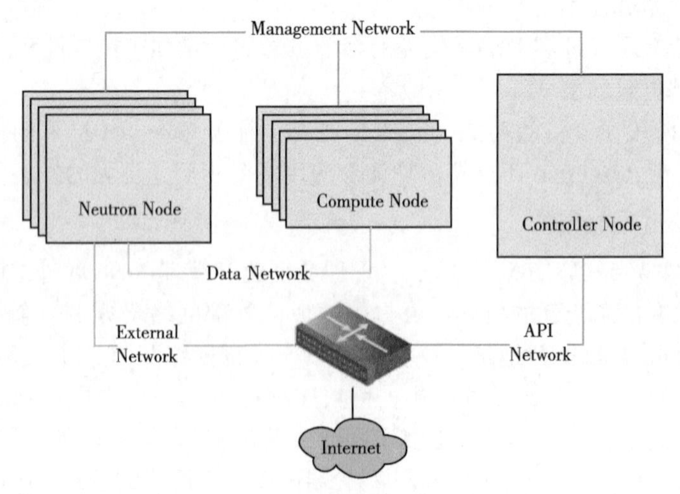

图 5－4　Neutron 网络类型

4. 网络模式

网络模式有 FlatDHCP 模式、VLAN 模式和 GRE 模式。

（1）FlatDHCP 模式：网桥模式，在网关处单独取了一个 DHCP 的进程，可以辅助用户进行网络配置。

FlatDHCP 模式通常指定一个子网，规定虚拟机能使用的 IP 地址范围，创建实例时从有效 IP 地址池获取一个 IP 为虚拟机实例分配，自动配置好网桥、通过 DNSmasq 分配地址，如图 5 –5 所示。

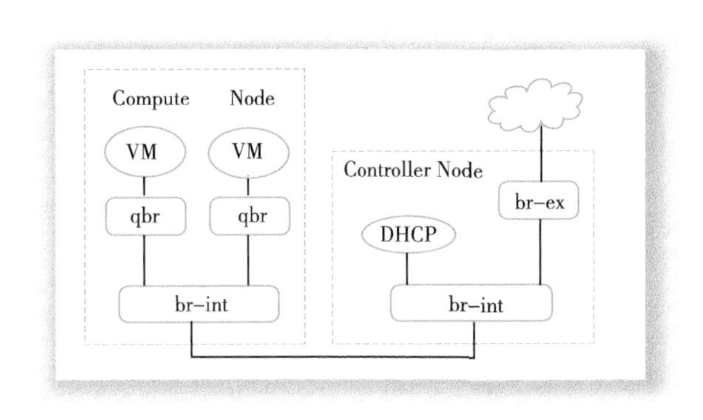

图 5 –5　FlatDHCP 模式

（2）VLAN 模式：为每个不同的租户设置了不同的虚拟子网，在这个虚拟子网中，用户可以有自己的 IP。

VLAN 模式需要创建租户 VLAN，使得租户之间两层网络隔离，并自动创建网桥，在网络控制器上的 DHCP 为所有的 VLAN 启动，为每个虚拟机分配私网地址（DNSmasq），网络控制器 NAT 转换，也解决两层隔离问题，适合私有云使用，如图 5 –6所示。

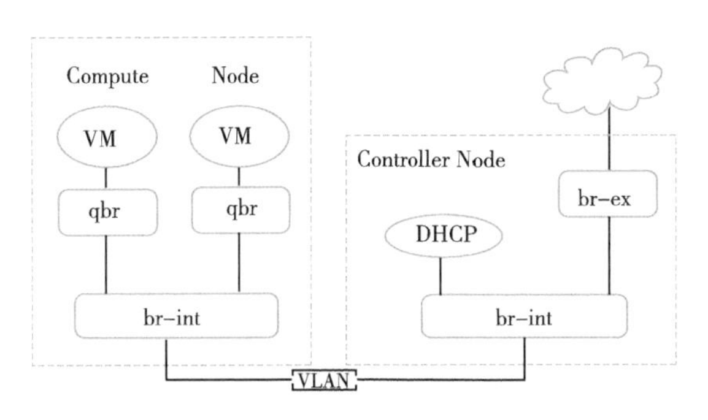

图 5 –6　VLAN 模式

（3）GRE 模式

GRE 网络实现了跨不同网络实现二次 IP 通信，而且通信封装在 IP 报文中，实现点对点隧道，如图 5 –7 所示。

在 OpenStack 中，所有网络有关的逻辑管理均在 Controller 节点中实现，如 DNS、DHCP 以及路由等。Compute 节点上只需要对所部属的虚拟机提供基本的网络功能支持，包括隔离不同租户的虚拟机和进行一些基本的安全策略管理（即 Security Group）。

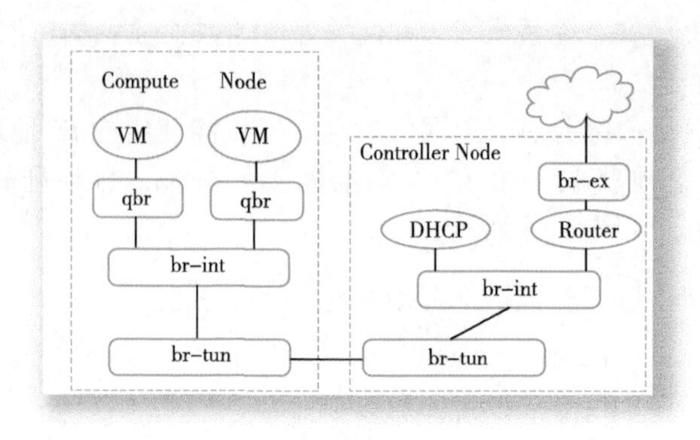

图 5－7　GRE 模式

视频
网络服务的基本操作

1. 准备实施任务的环境（见表 5－2）

表 5－2　实施任务的准备环境

实施任务所需软件资源	虚拟机镜像资源信息
虚拟机软件（vm12）	controller
securcrt 远程连接软件	compute

2. 分解实施任务的过程（见表 5－3）

表 5－3　实施任务的简明过程

序号	步骤	详细操作及说明
1	网络服务的基本操作	①列出系统扩展命令； ②修改网络模式； ③创建网络； ④创建子网； ⑤创建租户网络； ⑥创建用户子网； ⑦创建子网
2	网络隔离的基本操作	①创建安全组； ②安全组添加规则； ③查询安全组规则
3	创建符合企业需求的网络服务	①创建各部门项目； ②创建各部门项目和外来访问使用网络； ③网络隔离

3. 网络服务的基本操作

（1）列出系统扩展命令。

```
[root@ controller ~] # source /etc/keystone/admin - openrc. sh
[root@ controller ~] # neutron ext - list  - c alias  - c name
```

结果如图 5－8 所示。

```
[root@xiandian ~]# source /etc/keystone/admin-openrc.sh
[root@xiandian ~]# neutron ext-list -c alias -c name
+-----------------------------+-----------------------------------------------+
| alias                       | name                                          |
+-----------------------------+-----------------------------------------------+
| default-subnetpools         | Default Subnetpools                           |
| network-ip-availability     | Network IP Availability                       |
| network_availability_zone   | Network Availability Zone                     |
| auto-allocated-topology     | Auto Allocated Topology Services              |
| ext-gw-mode                 | Neutron L3 Configurable external gateway mode |
| binding                     | Port Binding                                  |
| agent                       | agent                                         |
| subnet_allocation           | Subnet Allocation                             |
| l3_agent_scheduler          | L3 Agent Scheduler                            |
| tag                         | Tag support                                   |
| external-net                | Neutron external network                      |
| net-mtu                     | Network MTU                                   |
| availability_zone           | Availability Zone                             |
| quotas                      | Quota management support                      |
| l3-ha                       | HA Router extension                           |
| provider                    | Provider Network                              |
| multi-provider              | Multi Provider Network                        |
| address-scope               | Address scope                                 |
| extraroute                  | Neutron Extra Route                           |
| timestamp_core              | Time Stamp Fields addition for core resources |
| router                      | Neutron L3 Router                             |
| extra_dhcp_opt              | Neutron Extra DHCP opts                       |
| dns-integration             | DNS Integration                               |
| security-group              | security-group                                |
| dhcp_agent_scheduler        | DHCP Agent Scheduler                          |
| router_availability_zone    | Router Availability Zone                      |
| rbac-policies               | RBAC Policies                                 |
| standard-attr-description   | standard-attr-description                     |
| port-security               | Port Security                                 |
| allowed-address-pairs       | Allowed Address Pairs                         |
| dvr                         | Distributed Virtual Router                    |
+-----------------------------+-----------------------------------------------+
```

图 5－8　扩展命令

（2）修改网络模式。

```
[root@ controller ~] #sed -i 101s/flat/vxlan/g'/etc/neutron/plugins/ml2/ml2_ conf. ini
[root@ controller ~] #crudini --set /etc/neutron/plugins/ml2/ml2_ conf. ini  ml2_ type_ vxlan vni_ ranges   1:1000
[root@ controller ~] #openstack -service restart
```

注：这个操作没有返回结果，重启的服务较多，等待时间需要两分钟左右。

（3）创建网络。

```
[root@ controller ~] #neutron net -create ext -net  --shared --router:external =True
```

结果如图 5－9 所示。

（4）创建子网。

```
[root@ controller ~] #neutron subnet -create ext -net --name ext -subnet \
  --allocation -pool start =172.24.7.100, end =172.24.7.200 \
  --disable -dhcp  --gateway 172.24.7.254 172.24.7.0/24
```

结果如图 5－10 所示。

```
[root@controller ~]# neutron net-create ext-net --shared --router:external=True
Created a new network:
+---------------------------+--------------------------------------+
| Field                     | Value                                |
+---------------------------+--------------------------------------+
| admin_state_up            | True                                 |
| availability_zone_hints   |                                      |
| availability_zones        |                                      |
| created_at                | 2019-11-19T00:38:21                  |
| description               |                                      |
| id                        | 27705173-b3b1-4dc0-8e4f-e01142000c3c |
| ipv4_address_scope        |                                      |
| ipv6_address_scope        |                                      |
| is_default                | False                                |
| mtu                       | 1458                                 |
| name                      | ext-net                              |
| port_security_enabled     | True                                 |
| provider:network_type     | gre                                  |
| provider:physical_network |                                      |
| provider:segmentation_id  | 36                                   |
| router:external           | True                                 |
| shared                    | True                                 |
| status                    | ACTIVE                               |
| subnets                   |                                      |
| tags                      |                                      |
| tenant_id                 | d2b88800df664a8cb196ec75034c9e45     |
| updated_at                | 2019-11-19T00:38:21                  |
+---------------------------+--------------------------------------+
```

图 5－9　创建网络

```
[root@controller ~]#  neutron subnet-create ext-net  --name ext-subnet \
> --allocation-pool start=172.24.7.100,end=172.24.7.200 \
> --disable-dhcp --gateway 172.24.7.254 172.24.7.0/24
Created a new subnet:
+-------------------+----------------------------------------------+
| Field             | Value                                        |
+-------------------+----------------------------------------------+
| allocation_pools  | {"start": "172.24.7.100", "end": "172.24.7.200"} |
| cidr              | 172.24.7.0/24                                |
| created_at        | 2019-11-19T00:39:23                          |
| description       |                                              |
| dns_nameservers   |                                              |
| enable_dhcp       | False                                        |
| gateway_ip        | 172.24.7.254                                 |
| host_routes       |                                              |
| id                | 044f1595-ad80-466a-8d96-8b01c9acc535         |
| ip_version        | 4                                            |
| ipv6_address_mode |                                              |
| ipv6_ra_mode      |                                              |
| name              | ext-subnet                                   |
| network_id        | 27705173-b3b1-4dc0-8e4f-e01142000c3c         |
| subnetpool_id     |                                              |
| tenant_id         | d2b88800df664a8cb196ec75034c9e45             |
| updated_at        | 2019-11-19T00:39:23                          |
+-------------------+----------------------------------------------+
```

图 5－10　创建子网

（5）创建租户网络。

```
[root@ controller ~]  # neutron net -create  demo -net
```

结果如图 5－11 所示。

（6）创建租户子网

```
[root@ controller ~]  # neutron subnet -create demo -net --
name demo -subnet   --gateway 10.0.0.1 10.0.0.0/24
```

结果如图 5－12 所示。

（7）创建路由。

```
[root@ controller ~]  # neutron router -create router1
```

结果如图 5－13 所示。

```
[root@xiandian ~]# neutron net-create  demo-net
Created a new network:
+---------------------------+--------------------------------------+
| Field                     | Value                                |
+---------------------------+--------------------------------------+
| admin_state_up            | True                                 |
| availability_zone_hints   |                                      |
| availability_zones        |                                      |
| created_at                | 2019-05-29T09:01:39                  |
| description               |                                      |
| id                        | b950a409-5fb2-48ed-9a7e-0af377b20546 |
| ipv4_address_scope        |                                      |
| ipv6_address_scope        |                                      |
| mtu                       | 1450                                 |
| name                      | demo-net                             |
| port_security_enabled     | True                                 |
| provider:network_type     | vxlan                                |
| provider:physical_network |                                      |
| provider:segmentation_id  | 80                                   |
| router:external           | False                                |
| shared                    | False                                |
| status                    | ACTIVE                               |
| subnets                   |                                      |
| tags                      |                                      |
| tenant_id                 | 0ab2dbde4f754b699e22461426cd0774     |
| updated_at                | 2019-05-29T09:01:39                  |
+---------------------------+--------------------------------------+
```

图 5－11　创建租户网络

```
[root@controller ~]# neutron subnet-create demo-net --name demo-subnet --gateway 10.0.0.1 10.0.0.0/
24
Created a new subnet:
+-------------------+----------------------------------------------+
| Field             | Value                                        |
+-------------------+----------------------------------------------+
| allocation_pools  | {"start": "10.0.0.2", "end": "10.0.0.254"}   |
| cidr              | 10.0.0.0/24                                  |
| created_at        | 2019-11-19T06:01:49                          |
| description       |                                              |
| dns_nameservers   |                                              |
| enable_dhcp       | True                                         |
| gateway_ip        | 10.0.0.1                                     |
| host_routes       |                                              |
| id                | 45a8aceb-e8ab-4d60-8886-317dee5a8163         |
| ip_version        | 4                                            |
| ipv6_address_mode |                                              |
| ipv6_ra_mode      |                                              |
| name              | demo-subnet                                  |
| network_id        | 806f6633-3d97-4d1f-823c-2123be4ab8d2         |
| subnetpool_id     |                                              |
| tenant_id         | d2b88800df664a8cb196ec75034c9e45             |
| updated_at        | 2019-11-19T06:01:49                          |
+-------------------+----------------------------------------------+
```

图 5－12　创建租户子网

```
[root@xiandian ~]# neutron router-create router1
Created a new router:
+-------------------------+--------------------------------------+
| Field                   | Value                                |
+-------------------------+--------------------------------------+
| admin_state_up          | True                                 |
| availability_zone_hints |                                      |
| availability_zones      |                                      |
| description             |                                      |
| distributed             | False                                |
| external_gateway_info   |                                      |
| ha                      | False                                |
| id                      | b266f80e-49e7-4332-822c-19ebd0ac1ce6 |
| name                    | router1                              |
| routes                  |                                      |
| status                  | ACTIVE                               |
| tenant_id               | 0ab2dbde4f754b699e22461426cd0774     |
+-------------------------+--------------------------------------+
```

图 5－13　创建路由

4．网络隔离的基本操作

（1）创建安全组。

```
[root@controller ~]# nova secgroup-create test "create a test secgroup"
```

结果如图 5 – 14 所示。

```
[root@xiandian ~]# nova secgroup-create test "create a test secgroup"
+--------------------------------------+------+------------------------+
| Id                                   | Name | Description            |
+--------------------------------------+------+------------------------+
| 83fbcaa8-a2c3-4176-837f-245d9f45cbd4 | test | create a test secgroup |
+--------------------------------------+------+------------------------+
```

图 5 – 14　创建安全组

（2）添加安全组规则。

```
[root@ controller ~] #nova secgroup -add -rule test ICMP -1 -1 172.24.4.0/24
[root@ controller ~] #nova secgroup -add -rule test TCP 1 65535 172.24.4.0/24
[root@ controller ~] #nova secgroup -add -rule test UDP 1 65535 172.24.4.0/24
[root@ controller ~] #nova secgroup -add -rule test ICMP -1 -1 172.24.5.0/24
[root@ controller ~] #nova secgroup -add -rule test TCP 1 65535 172.24.5.0/24
[root@ controller ~] #nova secgroup -add -rule test UDP 1 65535 172.24.5.0/24
```

结果如图 5 – 15 所示。

```
[root@xiandian ~]# nova secgroup-add-rule test ICMP -1 -1 172.24.4.0/24
+-------------+-----------+---------+---------------+--------------+
| IP Protocol | From Port | To Port | IP Range      | Source Group |
+-------------+-----------+---------+---------------+--------------+
| icmp        | -1        | -1      | 172.24.4.0/24 |              |
+-------------+-----------+---------+---------------+--------------+
[root@xiandian ~]# nova secgroup-add-rule test TCP 1 65535 172.24.4.0/24
+-------------+-----------+---------+---------------+--------------+
| IP Protocol | From Port | To Port | IP Range      | Source Group |
+-------------+-----------+---------+---------------+--------------+
| tcp         | 1         | 65535   | 172.24.4.0/24 |              |
+-------------+-----------+---------+---------------+--------------+
[root@xiandian ~]# nova secgroup-add-rule test UDP 1 65535 172.24.4.0/24
+-------------+-----------+---------+---------------+--------------+
| IP Protocol | From Port | To Port | IP Range      | Source Group |
+-------------+-----------+---------+---------------+--------------+
| udp         | 1         | 65535   | 172.24.4.0/24 |              |
+-------------+-----------+---------+---------------+--------------+
[root@xiandian ~]# nova secgroup-add-rule test ICMP -1 -1 172.24.5.0/24
+-------------+-----------+---------+---------------+--------------+
| IP Protocol | From Port | To Port | IP Range      | Source Group |
+-------------+-----------+---------+---------------+--------------+
| icmp        | -1        | -1      | 172.24.5.0/24 |              |
+-------------+-----------+---------+---------------+--------------+
[root@xiandian ~]# nova secgroup-add-rule test TCP 1 65535 172.24.5.0/24
+-------------+-----------+---------+---------------+--------------+
| IP Protocol | From Port | To Port | IP Range      | Source Group |
+-------------+-----------+---------+---------------+--------------+
| tcp         | 1         | 65535   | 172.24.5.0/24 |              |
+-------------+-----------+---------+---------------+--------------+
[root@xiandian ~]# nova secgroup-add-rule test UDP 1 65535 172.24.5.0/24
+-------------+-----------+---------+---------------+--------------+
| IP Protocol | From Port | To Port | IP Range      | Source Group |
+-------------+-----------+---------+---------------+--------------+
| udp         | 1         | 65535   | 172.24.5.0/24 |              |
+-------------+-----------+---------+---------------+--------------+
```

图 5 – 15　添加安全组规则

(3) 查询安全组规则。

[root@ controller ~] # nova secgroup - list - rules test

结果如图 5 - 16 所示。

```
[root@xiandian ~]# nova secgroup-list-rules test
+-------------+-----------+---------+---------------+--------------+
| IP Protocol | From Port | To Port | IP Range      | Source Group |
+-------------+-----------+---------+---------------+--------------+
| icmp        | -1        | -1      | 172.24.4.0/24 |              |
| udp         | 1         | 65535   | 172.24.4.0/24 |              |
| tcp         | 1         | 65535   | 172.24.4.0/24 |              |
| udp         | 1         | 65535   | 172.24.5.0/24 |              |
| tcp         | 1         | 65535   | 172.24.5.0/24 |              |
| icmp        | -1        | -1      | 172.24.5.0/24 |              |
+-------------+-----------+---------+---------------+--------------+
```

图 5 - 16　查询安全组规则

创建符合企业需求的网络服务

5. 创建符合企业需求的网络服务

(1) 通过 Dashboard 界面完成任务。

①创建各部门项目，如图 5 - 17 所示。

图 5 - 17　创建各部门项目

②创建各部门项目和外来访问使用网络。

a. 创建项目研发部项目的网络和子网，如图 5 - 18 和 5 - 19 所示。

b. 创建业务部项目的网络和子网，如图 5 - 20 和 5 - 21 所示。

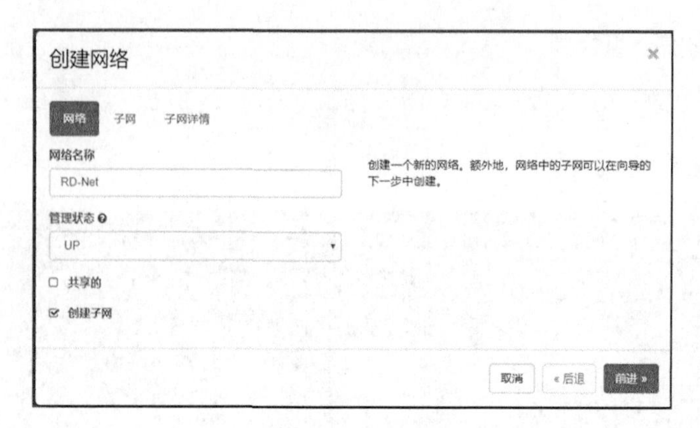

图 5－18　创建项目研发部项目的网络

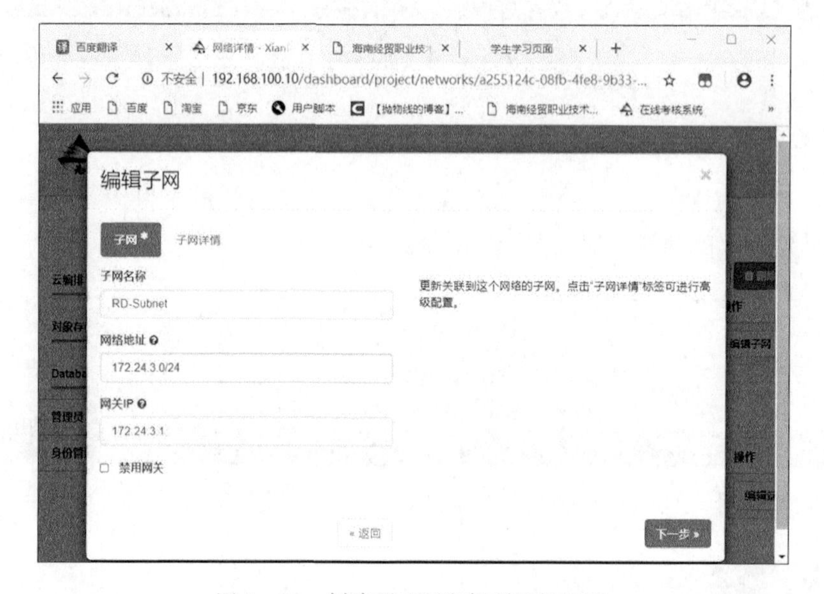

图 5－19　创建项目研发部项目的子网

图 5－20　创建业务部项目的网络

图 5－21　创建业务部项目的子网

c. 创建 IT 工程部项目的网络和子网，如图 5－22 和 5－23 所示。

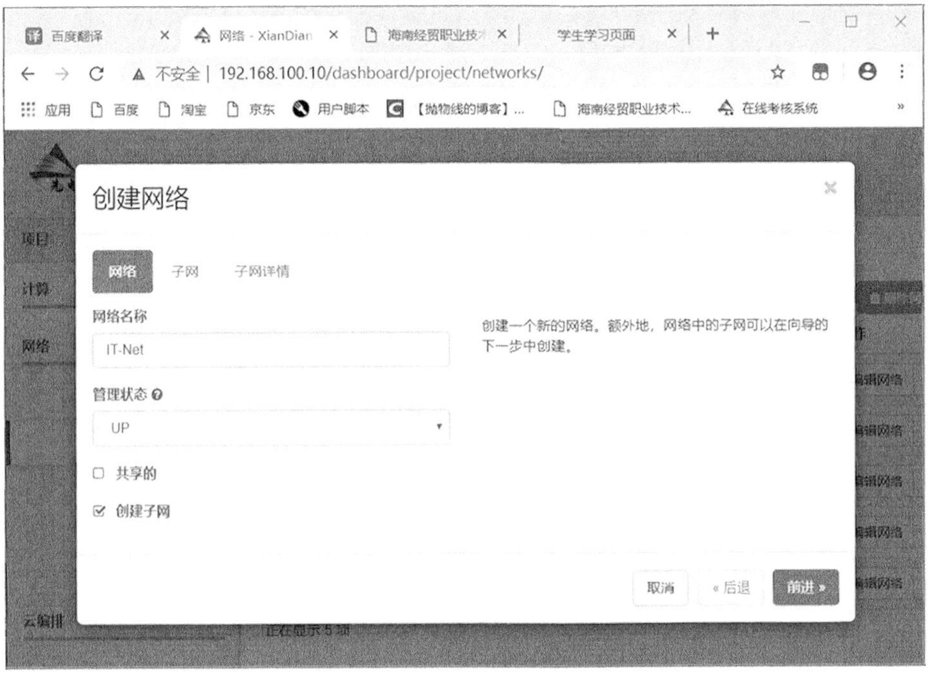

图 5－22　创建 IT 工程部项目的网络

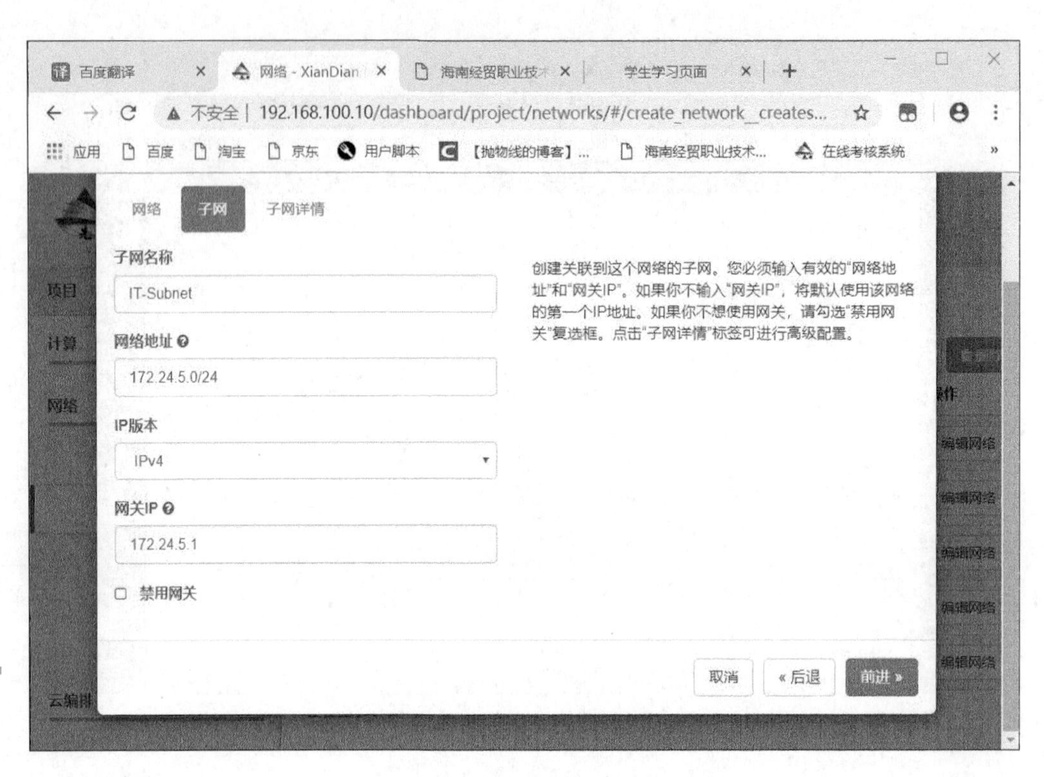

图 5－23　创建 IT 工程部项目的子网

d. 创建外来访问使用的网络和子网，如图 5－24 和 5－25 所示。

图 5－24　创建外来访问使用的网络

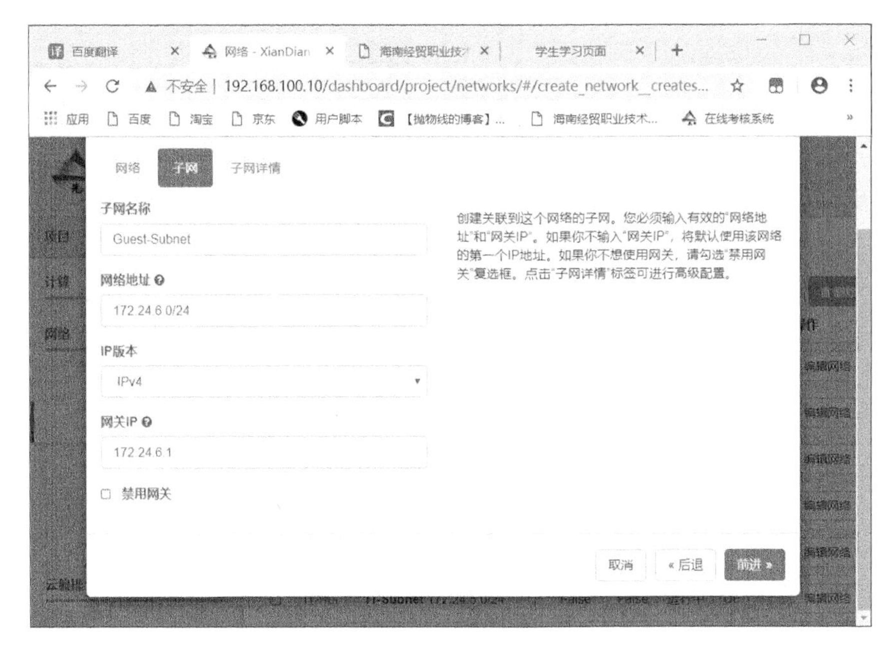

图 5-25 创建外来访问使用的子网

③网络隔离。

a. 创建项目研发部安全组规则，如图 5-26 所示。

创建安全组

名称 *

RD-Net

描述

说明：

安全组是IP过滤规则的集合，可被应用到虚拟机的网络设置中。在安全组创建后，你可以给安全组增加规则。

取消 创建安全组

图 5-26 创建项目研发部安全组规则

b. 创建业务部安全组规则，如图 5-27 所示。

c. 创建 IT 工程部安全组规则，如图 5-28 所示。

（2）通过 CLI 命令行完成任务。

①创建各部门项目，如图 5-29 所示。

```
[root@ controller ~] # openstack project create --domain = demo RD-Net
[root@ controller ~] # openstack project create --domain = demo BS-Net
[root@ controller ~] # openstack project create --domain = demo IT-Net
```

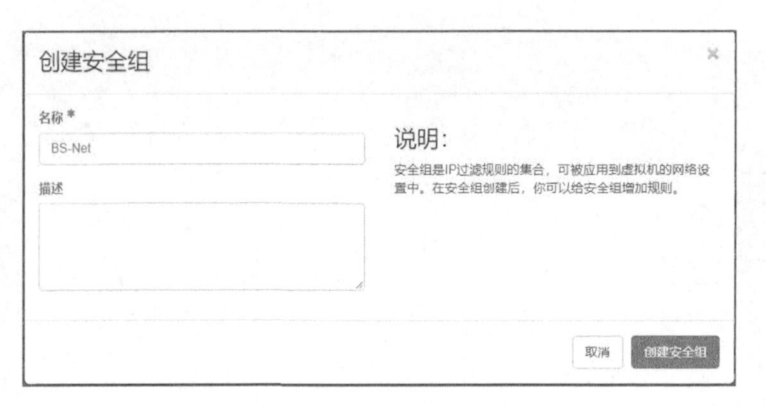

图 5－27　创建业务部安全组规则

创建安全组

名称 *

IT-Net

描述

说明：

安全组是IP过滤规则的集合，可被应用到虚拟机的网络设置中。在安全组创建后，你可以给安全组增加规则。

取消　创建安全组

图 5－28　创建 IT 工程部安全组规则

```
Neutron server returns request_ids: ['req-d0ce8eab-d3b8-44b2-aeb7-f522afbc2f8a']
[root@controller ~]# openstack project create --domain=demo RD-Net
+-------------+----------------------------------+
| Field       | Value                            |
+-------------+----------------------------------+
| description |                                  |
| domain_id   | 5b68813be9364902bf40477cb7f67d1d |
| enabled     | True                             |
| id          | aa81a9a792af4da3aa32ff3c214a0539 |
| is_domain   | False                            |
| name        | RD-Net                           |
| parent_id   | 5b68813be9364902bf40477cb7f67d1d |
+-------------+----------------------------------+
[root@controller ~]# openstack project  create --domain=demo BS-Net
+-------------+----------------------------------+
| Field       | Value                            |
+-------------+----------------------------------+
| description |                                  |
| domain_id   | 5b68813be9364902bf40477cb7f67d1d |
| enabled     | True                             |
| id          | 9c4e4532c02340b49fa72dbb0b42bbd8 |
| is_domain   | False                            |
| name        | BS-Net                           |
| parent_id   | 5b68813be9364902bf40477cb7f67d1d |
+-------------+----------------------------------+
[root@controller ~]# openstack project  create --domain=demo IT-Net
+-------------+----------------------------------+
| Field       | Value                            |
+-------------+----------------------------------+
| description |                                  |
| domain_id   | 5b68813be9364902bf40477cb7f67d1d |
| enabled     | True                             |
| id          | 3efb1087fa2345f6bd4639b797fe28a7 |
| is_domain   | False                            |
| name        | IT-Net                           |
| parent_id   | 5b68813be9364902bf40477cb7f67d1d |
+-------------+----------------------------------+
[root@controller ~]#
```

图 5－29　命令行创建各部门项目

②查看项目列表，如图 5－30 所示。

```
[root@ controller ~] #openstack project list
```

```
[root@controller ~]# openstack project list
+----------------------------------+---------+
| ID                               | Name    |
+----------------------------------+---------+
| 26b08dcad5a74c09b1ad0fda0a611bce | admin   |
| 3efb1087fa2345f6bd4639b797fe28a7 | IT-Net  |
| 52e83e9b3a8247a39a8af63e5ac445de | demo    |
| 6adff6539333430982aad1ae87256887 | service |
| 9c4e4532c02340b49fa72dbb0b42bbd8 | BS-Net  |
| aa81a9a792af4da3aa32ff3c214a0539 | RD-Net  |
+----------------------------------+---------+
[root@controller ~]#
```

图 5－30　查看项目列表

③创建各部门网络子网和外来访问使用的网络。

a. 为项目研发部创建网络和子网，如图 5－31 和图 5－32 所示。

```
[root@ controller ~] #openstack network create  --project projectID  RD-net
[root@ controller ~] #neutron subnet-create --name RD-Subnet --allocation-pool  start=172.24.3.100,end=172.24.3.200 --gateway 172.24.3.1  RD-net 172.24.3.0/24
```

```
[root@controller ~]# openstack network create  --project aa81a9a792af4da3aa32ff3c214a0539  RD-ne
t
+---------------------------+--------------------------------------+
| Field                     | Value                                |
+---------------------------+--------------------------------------+
| admin_state_up            | UP                                   |
| availability_zone_hints   |                                      |
| availability_zones        |                                      |
| created_at                | 2019-04-26T15:58:28                  |
| description               |                                      |
| headers                   |                                      |
| id                        | a04b3ff1-4741-4911-ae5e-61e2ddb96fed |
| ipv4_address_scope        | None                                 |
| ipv6_address_scope        | None                                 |
| mtu                       | 1458                                 |
| name                      | RD-net                               |
| port_security_enabled     | True                                 |
| project_id                | aa81a9a792af4da3aa32ff3c214a0539     |
| provider:network_type     | gre                                  |
| provider:physical_network | None                                 |
| provider:segmentation_id  | 21                                   |
| router_external           | Internal                             |
| shared                    | False                                |
| status                    | ACTIVE                               |
| subnets                   |                                      |
| tags                      | []                                   |
| updated_at                | 2019-04-26T15:58:28                  |
+---------------------------+--------------------------------------+
```

图 5－31　命令行创建项目研发部的网络和子网 1

```
[root@controller ~]# neutron subnet-create --name RD-Subnet --allocation-pool          star
t=172.24.3.100,end=172.24.3.200 --gateway 172.24.3.1  RD-net 172.24.3.0/24
Created a new subnet:
+-------------------+--------------------------------------------------+
| Field             | Value                                            |
+-------------------+--------------------------------------------------+
| allocation_pools  | {"start": "172.24.3.100", "end": "172.24.3.200"} |
| cidr              | 172.24.3.0/24                                    |
| created_at        | 2019-04-26T15:58:41                              |
| description       |                                                  |
| dns_nameservers   |                                                  |
| enable_dhcp       | True                                             |
| gateway_ip        | 172.24.3.1                                       |
| host_routes       |                                                  |
| id                | 31ac0558-8066-4686-addb-d084404cd5fb             |
| ip_version        | 4                                                |
| ipv6_address_mode |                                                  |
| ipv6_ra_mode      |                                                  |
| name              | RD-Subnet                                        |
| network_id        | a04b3ff1-4741-4911-ae5e-61e2ddb96fed             |
| subnetpool_id     |                                                  |
| tenant_id         | 26b08dcad5a74c09b1ad0fda0a611bce                 |
| updated_at        | 2019-04-26T15:58:41                              |
+-------------------+--------------------------------------------------+
[root@controller ~]#
```

图 5－32　命令行创建项目研发部的网络和子网 2

b. 为业务部创建网络和子网，如图 5－33 和图 5－34 所示。

```
[root@ controller ~] # openstack network create  --project projectID  BS-net
[root@ controller ~] #neutron subnet-create --name BS-Subnet --allocation-pool  start=172.24.4.100,end=172.24.4.200 --gateway 172.24.4.1  BS-Net 172.24.4.0/24
```

```
[root@controller ~]# openstack network create  --project BS-Net BS-net
+---------------------------+--------------------------------------+
| Field                     | Value                                |
+---------------------------+--------------------------------------+
| admin_state_up            | UP                                   |
| availability_zone_hints   |                                      |
| availability_zones        |                                      |
| created_at                | 2019-04-26T15:59:44                  |
| description               |                                      |
| headers                   |                                      |
| id                        | b965474c-dd2d-4c54-88c2-f8781e67e3ec |
| ipv4_address_scope        | None                                 |
| ipv6_address_scope        | None                                 |
| mtu                       | 1458                                 |
| name                      | BS-net                               |
| port_security_enabled     | True                                 |
| project_id                | 9c4e4532c02340b49fa72dbb0b42bbd8     |
| provider:network_type     | gre                                  |
| provider:physical_network | None                                 |
| provider:segmentation_id  | 30                                   |
| router_external           | Internal                             |
| shared                    | False                                |
| status                    | ACTIVE                               |
| subnets                   |                                      |
| tags                      | []                                   |
| updated_at                | 2019-04-26T15:59:44                  |
+---------------------------+--------------------------------------+
[root@controller ~]#
```

图 5－33　命令行创建业务部的网络和子网 1

```
[root@controller ~]# neutron subnet-create --name BS-Subnet --allocation-pool             star
t=172.24.4.100,end=172.24.4.200 --gateway 172.24.4.1  BS-Net 172.24.4.0/24
Created a new subnet:
+-------------------+--------------------------------------------------+
| Field             | Value                                            |
+-------------------+--------------------------------------------------+
| allocation_pools  | {"start": "172.24.4.100", "end": "172.24.4.200"} |
| cidr              | 172.24.4.0/24                                    |
| created_at        | 2019-04-26T16:00:07                              |
| description       |                                                  |
| dns_nameservers   |                                                  |
| enable_dhcp       | True                                             |
| gateway_ip        | 172.24.4.1                                       |
| host_routes       |                                                  |
| id                | 60c9df17-1d7d-4a80-956f-e6412bbc8f99             |
| ip_version        | 4                                                |
| ipv6_address_mode |                                                  |
| ipv6_ra_mode      |                                                  |
| name              | BS-Subnet                                        |
| network_id        | b965474c-dd2d-4c54-88c2-f8781e67e3ec             |
| subnetpool_id     |                                                  |
| tenant_id         | 26b08dcad5a74c09b1ad0fda0a611bce                 |
| updated_at        | 2019-04-26T16:00:07                              |
+-------------------+--------------------------------------------------+
[root@controller ~]#
```

图 5－34　命令行创建业务部的网络和子网 2

c. 为 IT 工程部通过命令创建网络和子网，如图 5－35 和 5－36 所示。

```
[root@ controller ~] # openstack network reate  --project projectID  IT-net
```

[root@ controller ~]# neutron subnet-create --name IT-Subnet --allocation-pool start=172.24.5.100,end=172.24.5.200 --gateway 172.24.5.1 IT-Net 172.24.5.0/24

```
+----------------------------------+---------+
| ID                               | Name    |
+----------------------------------+---------+
| 26b08dcad5a74c09b1ad0fda0a611bce | admin   |
| 3efb1087fa2345f6bd4639b797fe28a7 | IT-Net  |
| 52e83e9b3a8247a39a8af63e5ac445de | demo    |
| 6adff6539333430982aad1ae87256887 | service |
| 9c4e4532c02340b49fa72dbb0b42bbd8 | BS-Net  |
| aa81a9a792af4da3aa32ff3c214a0539 | RD-Net  |
+----------------------------------+---------+
[root@controller ~]# openstack network create  --project IT-Net  IT-net
+---------------------------+--------------------------------------+
| Field                     | Value                                |
+---------------------------+--------------------------------------+
| admin_state_up            | UP                                   |
| availability_zone_hints   |                                      |
| availability_zones        |                                      |
| created_at                | 2019-04-26T16:00:50                  |
| description               |                                      |
| headers                   |                                      |
| id                        | 381a3abc-8fb5-4bb3-8675-7b4ef9812037 |
| ipv4_address_scope        | None                                 |
| ipv6_address_scope        | None                                 |
| mtu                       | 1458                                 |
| name                      | IT-net                               |
| port_security_enabled     | True                                 |
| project_id                | 3efb1087fa2345f6bd4639b797fe28a7     |
| provider:network_type     | gre                                  |
| provider:physical_network | None                                 |
| provider:segmentation_id  | 79                                   |
| router_external           | Internal                             |
| shared                    | False                                |
| status                    | ACTIVE                               |
| subnets                   |                                      |
| tags                      | []                                   |
| updated_at                | 2019-04-26T16:00:50                  |
+---------------------------+--------------------------------------+
[root@controller ~]#
```

图 5-35　命令行创建 IT 工程部的网络和子网 1

```
| name                      | IT-net                               |
| port_security_enabled     | True                                 |
| project_id                | 3efb1087fa2345f6bd4639b797fe28a7     |
| provider:network_type     | gre                                  |
| provider:physical_network | None                                 |
| provider:segmentation_id  | 79                                   |
| router_external           | Internal                             |
| shared                    | False                                |
| status                    | ACTIVE                               |
| subnets                   |                                      |
| tags                      | []                                   |
| updated_at                | 2019-04-26T16:00:50                  |
+---------------------------+--------------------------------------+
[root@controller ~]# neutron subnet-create --name IT-Subnet --allocation-pool star
t=172.24.5.100,end=172.24.5.200 --gateway 172.24.5.1  IT-Net 172.24.5.0/24
Created a new subnet:
+-------------------+--------------------------------------------------+
| Field             | Value                                            |
+-------------------+--------------------------------------------------+
| allocation_pools  | {"start": "172.24.5.100", "end": "172.24.5.200"} |
| cidr              | 172.24.5.0/24                                    |
| created_at        | 2019-04-26T16:01:12                              |
| description       |                                                  |
| dns_nameservers   |                                                  |
| enable_dhcp       | True                                             |
| gateway_ip        | 172.24.5.1                                       |
| host_routes       |                                                  |
| id                | ebc5f0b6-a2fd-40f7-8ffa-1fd94ef1e638             |
| ip_version        | 4                                                |
| ipv6_address_mode |                                                  |
| ipv6_ra_mode      |                                                  |
| name              | IT-Subnet                                        |
| network_id        | 381a3abc-8fb5-4bb3-8675-7b4ef9812037             |
| subnetpool_id     |                                                  |
| tenant_id         | 26b08dcad5a74c09b1ad0fda0a611bce                 |
| updated_at        | 2019-04-26T16:01:12                              |
+-------------------+--------------------------------------------------+
[root@controller ~]#
```

图 5-36　命令行创建 IT 工程部的网络和子网 2

d. 创建外来访问使用的网络和子网，如图 5-37 和 5-38 所示。

[root@ controller ~] # openstack network create Guest -Net

[root@ controller ~] # neutron subnet -create --name Guest -Subnet --allocation -pool start =172.24.6.100,end = 172.24.6.200 --gateway 172.24.6.1 Guest -Net 172.24.6.0/24

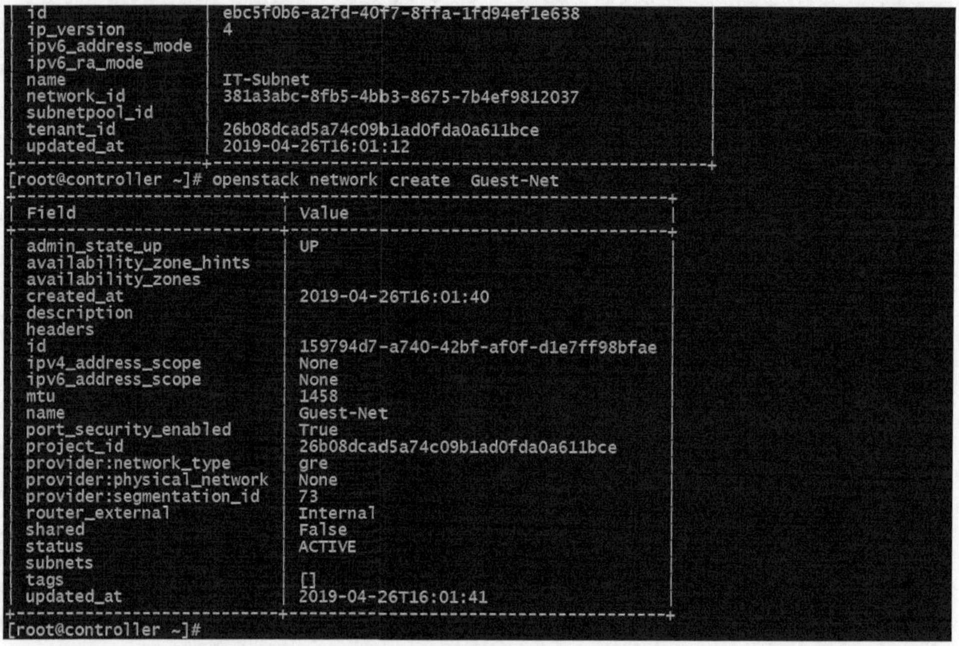

```
| id                | ebc5f0b6-a2fd-40f7-8ffa-1fd94ef1e638 |
| ip_version        | 4                                    |
| ipv6_address_mode |                                      |
| ipv6_ra_mode      |                                      |
| name              | IT-Subnet                            |
| network_id        | 381a3abc-8fb5-4bb3-8675-7b4ef9812037 |
| subnetpool_id     |                                      |
| tenant_id         | 26b08dcad5a74c09b1ad0fda0a611bce     |
| updated_at        | 2019-04-26T16:01:12                  |
+-------------------+--------------------------------------+
[root@controller ~]# openstack network create  Guest-Net
+---------------------------+--------------------------------------+
| Field                     | Value                                |
+---------------------------+--------------------------------------+
| admin_state_up            | UP                                   |
| availability_zone_hints   |                                      |
| availability_zones        |                                      |
| created_at                | 2019-04-26T16:01:40                  |
| description               |                                      |
| headers                   |                                      |
| id                        | 159794d7-a740-42bf-af0f-d1e7ff98bfae |
| ipv4_address_scope        | None                                 |
| ipv6_address_scope        | None                                 |
| mtu                       | 1458                                 |
| name                      | Guest-Net                            |
| port_security_enabled     | True                                 |
| project_id                | 26b08dcad5a74c09b1ad0fda0a611bce     |
| provider:network_type     | gre                                  |
| provider:physical_network | None                                 |
| provider:segmentation_id  | 73                                   |
| router_external           | Internal                             |
| shared                    | False                                |
| status                    | ACTIVE                               |
| subnets                   |                                      |
| tags                      | []                                   |
| updated_at                | 2019-04-26T16:01:41                  |
+---------------------------+--------------------------------------+
[root@controller ~]#
```

图 5－37　命令行创建外来访问使用的网络和子网 1

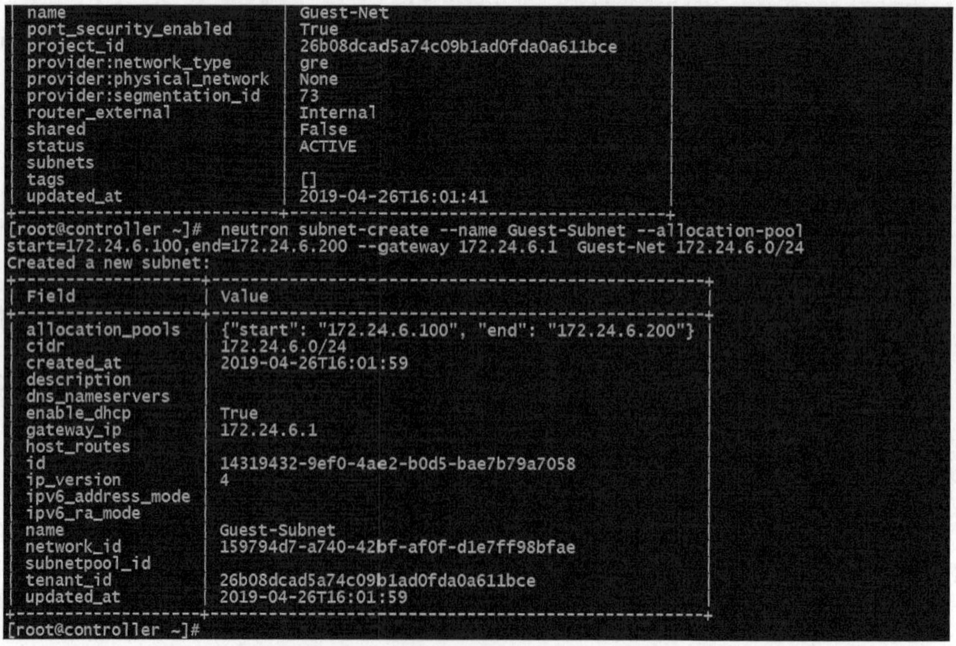

```
| name                      | Guest-Net                            |
| port_security_enabled     | True                                 |
| project_id                | 26b08dcad5a74c09b1ad0fda0a611bce     |
| provider:network_type     | gre                                  |
| provider:physical_network | None                                 |
| provider:segmentation_id  | 73                                   |
| router_external           | Internal                             |
| shared                    | False                                |
| status                    | ACTIVE                               |
| subnets                   |                                      |
| tags                      | []                                   |
| updated_at                | 2019-04-26T16:01:41                  |
+---------------------------+--------------------------------------+
[root@controller ~]#  neutron subnet-create --name Guest-Subnet --allocation-pool
start=172.24.6.100,end=172.24.6.200 --gateway 172.24.6.1  Guest-Net 172.24.6.0/24
Created a new subnet:
+-------------------+----------------------------------------------------+
| Field             | Value                                              |
+-------------------+----------------------------------------------------+
| allocation_pools  | {"start": "172.24.6.100", "end": "172.24.6.200"}   |
| cidr              | 172.24.6.0/24                                      |
| created_at        | 2019-04-26T16:01:59                                |
| description       |                                                    |
| dns_nameservers   |                                                    |
| enable_dhcp       | True                                               |
| gateway_ip        | 172.24.6.1                                         |
| host_routes       |                                                    |
| id                | 14319432-9ef0-4ae2-b0d5-bae7b79a7058               |
| ip_version        | 4                                                  |
| ipv6_address_mode |                                                    |
| ipv6_ra_mode      |                                                    |
| name              | Guest-Subnet                                       |
| network_id        | 159794d7-a740-42bf-af0f-d1e7ff98bfae               |
| subnetpool_id     |                                                    |
| tenant_id         | 26b08dcad5a74c09b1ad0fda0a611bce                   |
| updated_at        | 2019-04-26T16:01:59                                |
+-------------------+----------------------------------------------------+
[root@controller ~]#
```

图 5－38　命令行创建外来访问使用的网络和子网 2

e. 查看网络和子网，如图 5－39 所示。

```
[root@ controller ~] #openstack network list
[root@ controller ~] #openstack subnet list
```

```
| name              | Guest-Subnet                         |
| network_id        | 159794d7-a740-42bf-af0f-d1e7ff98bfae |
| subnetpool_id     |                                      |
| tenant_id         | 26b08dcad5a74c09b1ad0fda0a611bce     |
| updated_at        | 2019-04-26T16:01:59                  |
+-------------------+--------------------------------------+
[root@controller ~]#  openstack network list
+--------------------------------------+------------+--------------------------------------+
| ID                                   | Name       | Subnets                              |
+--------------------------------------+------------+--------------------------------------+
| 64e67498-e167-4291-9f20-6f3185c7e6ef | ext-net    | a3d8f61b-766a-4637-85fb-a88b1aab2997 |
| 0c80beec-b901-4cef-909c-89ab12427383 | int-net1   | 338439fa-579f-486c-8a64-2babc7fbb1aa |
| 159794d7-a740-42bf-af0f-d1e7ff98bfae | Guest-Net  | 14319432-9ef0-4ae2-b0d5-bae7b79a7058 |
| 381a3abc-8fb5-4bb3-8675-7b4ef9812037 | IT-net     | ebc5f0b6-a2fd-40f7-8ffa-1fd94ef1e638 |
| 3c79fef2-4a04-448a-97e3-2316ed7171ef | int-net2   | 3d6276ad-9e9a-4657-b2e9-92510d4e0913 |
| a04b3ff1-4741-4911-ae5e-61e2ddb96fed | RD-net     | 31ac0558-8066-4686-addb-d084404cd5fb |
| b965474c-dd2d-4c54-88c2-f8781e67e3ec | BS-net     | 60c9df17-1d7d-4a80-956f-e6412bbc8f99 |
+--------------------------------------+------------+--------------------------------------+
[root@controller ~]# openstack subnet list
+--------------------------+---------------+---------------------------+------------------+
| ID                       | Name          | Network                   | Subnet           |
+--------------------------+---------------+---------------------------+------------------+
| a3d8f61b-766a-4637-85fb- | ext-subnet    | 64e67498-e167-4291-9f20-6f31 | 192.168.200.0/24 |
| a88b1aab2997             |               | 85c7e6ef                  |                  |
| 14319432-9ef0-4ae2-b0d5-bae | Guest-Subnet | 159794d7-a740-42bf-af0f- | 172.24.6.0/24    |
| 7b79a7058                |               | d1e7ff98bfae              |                  |
| 31ac0558-8066-4686-addb- | RD-Subnet     | a04b3ff1-4741-4911-ae5e-  | 172.24.3.0/24    |
| d084404cd5fb             |               | 61e2ddb96fed              |                  |
| 338439fa-579f-486c-      | int-subnet1   | 0c80beec-b901-4cef-909c-  | 10.0.0.0/24      |
| 8a64-2babc7fbb1aa        |               | 89ab12427383              |                  |
| 3d6276ad-                | int-subnet2   | 3c79fef2-4a04-448a-       | 10.0.1.0/24      |
| 9e9a-4657-b2e9-92510d4e0913 |            | 97e3-2316ed7171ef         |                  |
| 60c9df17-1d7d-4a80-956f- | BS-Subnet     | b965474c-dd2d-            | 172.24.4.0/24    |
| e6412bbc8f99             |               | 4c54-88c2-f8781e67e3ec    |                  |
| ebc5f0b6-a2fd-40f7-8ffa- | IT-Subnet     | 381a3abc-                 | 172.24.5.0/24    |
| 1fd94ef1e638             |               | 8fb5-4bb3-8675-7b4ef9812037 |                |
+--------------------------+---------------+---------------------------+------------------+
[root@controller ~]#
```

图 5－39　命令行查看网络和子网

④网络隔离。

a. 创建项目研发部安全组规则，如图 5－40 所示。

```
[root@ controller ~]  #nova secgroup-create RD_ Rule RD
[root@ controller ~]  #nova secgroup-add-rule RD_Rule ICMP -1 -1 172.24.3.0/24
[root@ controller ~]  #nova secgroup-add-rule RD_Rule  TCP 1  65535 172.24.3.0/24
[root@ controller ~]  #nova secgroup-add-rule RD_Rule  UDP 1  65535 172.24.3.0/24
```

```
[root@controller ~]# nova secgroup-create RD_Rule RD
+--------------------------------------+---------+-------------+
| Id                                   | Name    | Description |
+--------------------------------------+---------+-------------+
| 05dba2de-c1ce-4df1-bbf8-900eed803854 | RD_Rule | RD          |
+--------------------------------------+---------+-------------+
[root@controller ~]# nova secgroup-add-rule RD_Rule ICMP -1 -1 172.24.3.0/24
+-------------+-----------+---------+---------------+--------------+
| IP Protocol | From Port | To Port | IP Range      | Source Group |
+-------------+-----------+---------+---------------+--------------+
| icmp        | -1        | -1      | 172.24.3.0/24 |              |
+-------------+-----------+---------+---------------+--------------+
[root@controller ~]# nova secgroup-add-rule RD_Rule  TCP  1  65535 172.24.3.0/24
+-------------+-----------+---------+---------------+--------------+
| IP Protocol | From Port | To Port | IP Range      | Source Group |
+-------------+-----------+---------+---------------+--------------+
| tcp         | 1         | 65535   | 172.24.3.0/24 |              |
+-------------+-----------+---------+---------------+--------------+
[root@controller ~]# nova secgroup-add-rule RD_Rule  UDP  1  65535 172.24.3.0/24
+-------------+-----------+---------+---------------+--------------+
| IP Protocol | From Port | To Port | IP Range      | Source Group |
+-------------+-----------+---------+---------------+--------------+
| udp         | 1         | 65535   | 172.24.3.0/24 |              |
+-------------+-----------+---------+---------------+--------------+
[root@controller ~]#
```

图 5－40　命令行创建项目研发部安全组规则

项目研发部与 IT 工程部可以相互访问，如图 5－41 所示。

```
[root@ controller ~]# nova secgroup - add - rule RD_Rule ICMP -1 -1 172.24.5.0/24
[root@ controller ~]# nova secgroup - add - rule RD_Rule  TCP 1  65535 172.24.5.0/24
[root@ controller ~]# nova secgroup - add - rule RD_Rule  UDP 1  65535 172.24.5.0/24
```

```
[root@controller ~]# nova secgroup-add-rule RD_Rule ICMP -1 -1 172.24.5.0/24
+-------------+-----------+---------+---------------+--------------+
| IP Protocol | From Port | To Port | IP Range      | Source Group |
+-------------+-----------+---------+---------------+--------------+
| icmp        | -1        | -1      | 172.24.5.0/24 |              |
+-------------+-----------+---------+---------------+--------------+
[root@controller ~]# nova secgroup-add-rule RD_Rule  TCP   1  65535 172.24.5.0/24
+-------------+-----------+---------+---------------+--------------+
| IP Protocol | From Port | To Port | IP Range      | Source Group |
+-------------+-----------+---------+---------------+--------------+
| tcp         | 1         | 65535   | 172.24.5.0/24 |              |
+-------------+-----------+---------+---------------+--------------+
[root@controller ~]# nova secgroup-add-rule RD_Rule  UDP   1  65535 172.24.5.0/24
+-------------+-----------+---------+---------------+--------------+
| IP Protocol | From Port | To Port | IP Range      | Source Group |
+-------------+-----------+---------+---------------+--------------+
| udp         | 1         | 65535   | 172.24.5.0/24 |              |
+-------------+-----------+---------+---------------+--------------+
[root@controller ~]#
```

图 5－41　命令行使得项目研发部与 IT 工程部相互访问

查看安全组规则，如图 5－42 所示。

```
[root@ controller ~] #nova secgroup - list - rules RD_ Rule
```

```
[root@controller ~]# nova secgroup-list-rules RD_Rule
+-------------+-----------+---------+---------------+--------------+
| IP Protocol | From Port | To Port | IP Range      | Source Group |
+-------------+-----------+---------+---------------+--------------+
| icmp        | -1        | -1      | 172.24.5.0/24 |              |
| icmp        | -1        | -1      | 172.24.3.0/24 |              |
| tcp         | 1         | 65535   | 172.24.3.0/24 |              |
| udp         | 1         | 65535   | 172.24.5.0/24 |              |
| udp         | 1         | 65535   | 172.24.3.0/24 |              |
| tcp         | 1         | 65535   | 172.24.5.0/24 |              |
+-------------+-----------+---------+---------------+--------------+
[root@controller ~]#
```

图 5－42　命令行查看安全组规则

b. 创建业务部安全组规则，如图 5－43 所示。

```
[root@ controller ~]# nova secgroup - create BS_Rule BS
[root@ controller ~]# nova secgroup - add - rule BS_Rule ICMP -1  -1 172.24.4.0/24
[root@ controller ~]# nova secgroup - add - rule BS_Rule  TCP 1  65535 172.24.4.0/24
[root@ controller ~]# nova secgroup - add - rule BS_Rule  UCP 1  65535 172.24.4.0/24
```

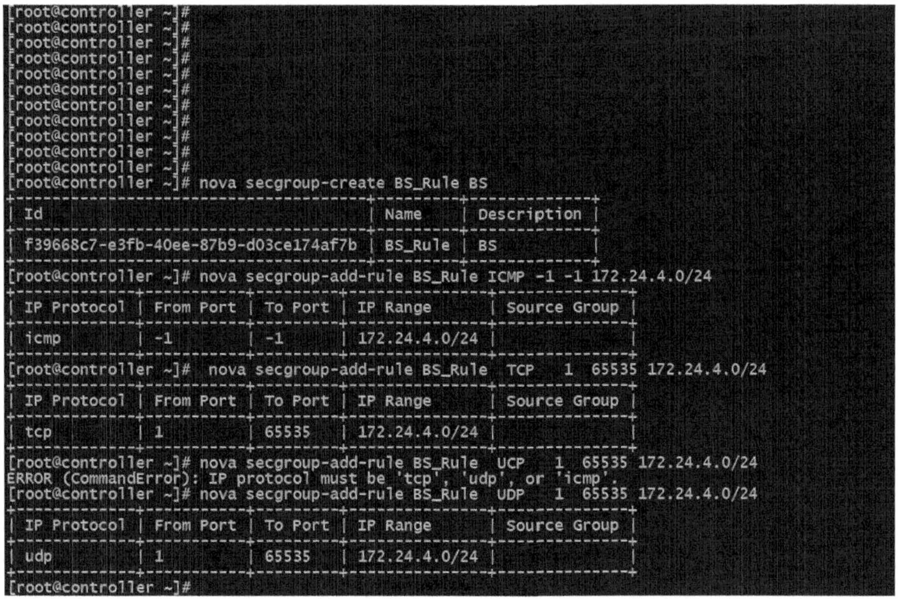

```
[root@controller ~]#
[root@controller ~]#
[root@controller ~]#
[root@controller ~]#
[root@controller ~]#
[root@controller ~]#
[root@controller ~]#
[root@controller ~]#
[root@controller ~]#
[root@controller ~]#
[root@controller ~]#
[root@controller ~]# nova secgroup-create BS_Rule BS
+--------------------------------------+---------+-------------+
| Id                                   | Name    | Description |
+--------------------------------------+---------+-------------+
| f39668c7-e3fb-40ee-87b9-d03ce174af7b | BS_Rule | BS          |
+--------------------------------------+---------+-------------+
[root@controller ~]# nova secgroup-add-rule BS_Rule ICMP -1 -1 172.24.4.0/24
+-------------+-----------+---------+---------------+--------------+
| IP Protocol | From Port | To Port | IP Range      | Source Group |
+-------------+-----------+---------+---------------+--------------+
| icmp        | -1        | -1      | 172.24.4.0/24 |              |
+-------------+-----------+---------+---------------+--------------+
[root@controller ~]#  nova secgroup-add-rule BS_Rule  TCP   1  65535 172.24.4.0/24
+-------------+-----------+---------+---------------+--------------+
| IP Protocol | From Port | To Port | IP Range      | Source Group |
+-------------+-----------+---------+---------------+--------------+
| tcp         | 1         | 65535   | 172.24.4.0/24 |              |
+-------------+-----------+---------+---------------+--------------+
[root@controller ~]# nova secgroup-add-rule BS_Rule  UCP   1  65535 172.24.4.0/24
ERROR (CommandError): IP protocol must be 'tcp', 'udp', or 'icmp'.
[root@controller ~]# nova secgroup-add-rule BS_Rule  UDP   1  65535 172.24.4.0/24
+-------------+-----------+---------+---------------+--------------+
| IP Protocol | From Port | To Port | IP Range      | Source Group |
+-------------+-----------+---------+---------------+--------------+
| udp         | 1         | 65535   | 172.24.4.0/24 |              |
+-------------+-----------+---------+---------------+--------------+
[root@controller ~]#
```

图 5－43　命令行创建业务部安全组规则

业务部与 IT 工程部可以相互访问，如图 5－44 所示。

```
[root@ controller ~]# nova secgroup - add - rule BS_Rule ICMP
-1 -1 172.24.5.0/24
[root@ controller ~]# nova secgroup - add - rule BS_Rule  TCP
1  65535 172.24.5.0/24
[root@ controller ~]# nova secgroup - add - rule BS_Rule  UDP
1  65535 172.24.5.0/24
```

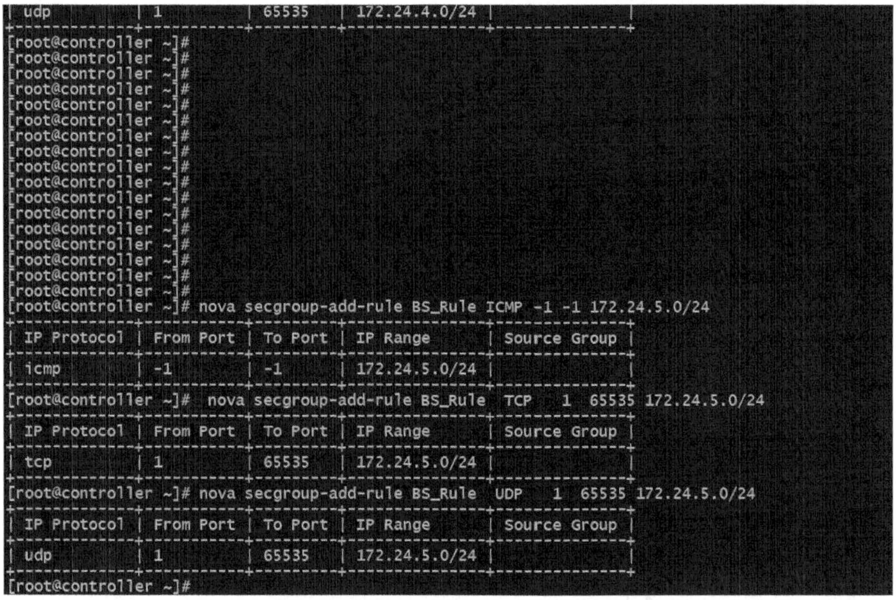

```
| udp         | 1         | 65535   | 172.24.4.0/24 |              |
+-------------+-----------+---------+---------------+--------------+
[root@controller ~]#
[root@controller ~]#
[root@controller ~]#
[root@controller ~]#
[root@controller ~]#
[root@controller ~]#
[root@controller ~]#
[root@controller ~]#
[root@controller ~]#
[root@controller ~]#
[root@controller ~]#
[root@controller ~]#
[root@controller ~]#
[root@controller ~]#
[root@controller ~]#
[root@controller ~]#
[root@controller ~]#
[root@controller ~]# nova secgroup-add-rule BS_Rule ICMP -1 -1 172.24.5.0/24
+-------------+-----------+---------+---------------+--------------+
| IP Protocol | From Port | To Port | IP Range      | Source Group |
+-------------+-----------+---------+---------------+--------------+
| icmp        | -1        | -1      | 172.24.5.0/24 |              |
+-------------+-----------+---------+---------------+--------------+
[root@controller ~]#  nova secgroup-add-rule BS_Rule  TCP   1  65535 172.24.5.0/24
+-------------+-----------+---------+---------------+--------------+
| IP Protocol | From Port | To Port | IP Range      | Source Group |
+-------------+-----------+---------+---------------+--------------+
| tcp         | 1         | 65535   | 172.24.5.0/24 |              |
+-------------+-----------+---------+---------------+--------------+
[root@controller ~]# nova secgroup-add-rule BS_Rule  UDP   1  65535 172.24.5.0/24
+-------------+-----------+---------+---------------+--------------+
| IP Protocol | From Port | To Port | IP Range      | Source Group |
+-------------+-----------+---------+---------------+--------------+
| udp         | 1         | 65535   | 172.24.5.0/24 |              |
+-------------+-----------+---------+---------------+--------------+
[root@controller ~]#
```

图 5－44　命令行使得业务部与 IT 工程部相互访问

查看安全组规则，如图 5－45 所示。

```
[root@ controller ~] #nova secgroup -list -rules BS_ Rule
```

```
[root@controller ~]#
[root@controller ~]# nova secgroup-list-rules BS_Rule
+-------------+-----------+---------+---------------+--------------+
| IP Protocol | From Port | To Port | IP Range      | Source Group |
+-------------+-----------+---------+---------------+--------------+
| icmp        | -1        | -1      | 172.24.4.0/24 |              |
| udp         | 1         | 65535   | 172.24.4.0/24 |              |
| udp         | 1         | 65535   | 172.24.5.0/24 |              |
| tcp         | 1         | 65535   | 172.24.5.0/24 |              |
| icmp        | -1        | -1      | 172.24.5.0/24 |              |
| tcp         | 1         | 65535   | 172.24.4.0/24 |              |
+-------------+-----------+---------+---------------+--------------+
[root@controller ~]#
```

图 5－45　命令行查看安全组规则

c. 创建 IT 工程部安全组规则，如图 5－46 所示。

```
[root@ controller ~]  #nova secgroup -create IT_Rule  IT
[root@ controller ~]  #nova secgroup -add -rule  IT_Rule
ICMP -1 -1 172.24.5.0/24
[root@ controller ~]  #nova secgroup -add -rule  IT_Rule  TCP
1  65535 172.24.5.0/24
[root@ controller ~]  #nova secgroup -add -rule  IT_Rule  UDP
1  65535 172.24.5.0/24
```

```
[root@controller ~]#
[root@controller ~]# nova secgroup-create IT_Rule  IT
+--------------------------------------+---------+-------------+
| Id                                   | Name    | Description |
+--------------------------------------+---------+-------------+
| b983d498-0e3e-4bc7-9107-6ce88e9abac3 | IT_Rule | IT          |
+--------------------------------------+---------+-------------+
[root@controller ~]# nova secgroup-add-rule  IT_Rule ICMP -1 -1 172.24.5.0/24
+-------------+-----------+---------+---------------+--------------+
| IP Protocol | From Port | To Port | IP Range      | Source Group |
+-------------+-----------+---------+---------------+--------------+
| icmp        | -1        | -1      | 172.24.5.0/24 |              |
+-------------+-----------+---------+---------------+--------------+
[root@controller ~]# nova secgroup-add-rule  IT_Rule  TCP   1  65535 172.24.5.0/24
+-------------+-----------+---------+---------------+--------------+
| IP Protocol | From Port | To Port | IP Range      | Source Group |
+-------------+-----------+---------+---------------+--------------+
| tcp         | 1         | 65535   | 172.24.5.0/24 |              |
+-------------+-----------+---------+---------------+--------------+
[root@controller ~]# nova secgroup-add-rule  IT_Rule  UDP   1  65535 172.24.5.0/24
+-------------+-----------+---------+---------------+--------------+
| IP Protocol | From Port | To Port | IP Range      | Source Group |
+-------------+-----------+---------+---------------+--------------+
| udp         | 1         | 65535   | 172.24.5.0/24 |              |
+-------------+-----------+---------+---------------+--------------+
[root@controller ~]#
```

图 5－46　命令行创建 IT 工程部安全组规则

查看安全组规则，如图 5－47 所示。

```
[root@ controller ~] #nova secgroup -list -rules IT_ Rule
```

```
[root@controller ~]#
[root@controller ~]#  nova secgroup-list-rules IT_Rule
+-------------+-----------+---------+---------------+--------------+
| IP Protocol | From Port | To Port | IP Range      | Source Group |
+-------------+-----------+---------+---------------+--------------+
| tcp         | 1         | 65535   | 172.24.5.0/24 |              |
| udp         | 1         | 65535   | 172.24.5.0/24 |              |
| icmp        | -1        | -1      | 172.24.5.0/24 |              |
+-------------+-----------+---------+---------------+--------------+
[root@controller ~]#
```

图 5－47　命令行安全组规则

项目小结

传统的网络管理方式很大程度依赖于管理员手工配置和维护各种网络硬件设备，而云环境下的网络已经变得非常复杂，“软件定义网络（SDN）”所具有的灵活性和自动化优势使其成为云时代网络管理的主流，OpenStack 的 Neutron 就是实现 SDN 的功能，本项目讨论 Neutron 功能和它的各个组件及运行机制，通过学习网络服务的基本操作及根据企业需求部署和配置 Neutron 网络任务，使读者较好地完成项目目标。

项目扩展

表 5 –4 列举本项目 Neutron 的常用命令及其说明。

表 5 –4　常用命令及其说明

常用命令	命令说明
neutron net – list	列出当前租户所有的网络
neutron net – list　– – all – tenants	列出所有租户的所有网络
neutron net – create	创建一个网络
neutron net – show NET_ ID	查看一个网络的详细信息
neutron net – delete NET_ ID	删除一个网络
neutron subnet – create	创建一个子网
neutron subnet – delete	删除一个子网
neutron subnet – list	显示子网
neutron subnet – show	查看一个子网的详细信息
neutron router – create	创建一个路由
neutron router – delete	删除一个路由
neutron router – update	更新路由的信息
neutron router – show	查看路由的信息
neutron router – list	列出所有的路由

项目思考

根据本项目的介绍，Neutron 组件支持哪些网络类型？

项目六

计算服务

项目综述

小张已经大致了解到计算服务对整个云计算平台的重要性，他还要深入了解计算服务 Nova 的具体作用和实现的功能，首先需要了解什么是 Nova 及其底层的运行机制和原理，了解底层调用的虚拟化组件，明白与其他组件的联系。最终要掌握的技能是完成在 Dashboard 界面及以命令行正常启动、关闭和重建虚拟机等操作。小张为此要完成的任务如下：

项目六综述

- 计算服务的基本操作。
- 在 Dashboard 界面上启动和部署云主机，并进行测试。

项目目标

【知识目标】
- OpenStack 的 Nova 服务组件的相关概念。
- OpenStack 的 Nova 组件的服务流程和工作机制。

【技能目标】
- 计算服务的基本操作。
- 在 Dashboard 界面上启动和部署云主机，并进行测试。

【职业能力】

根据企业的应用需求，正常启动、关闭和重建云主机等操作。

任务　安装与操作计算服务

任务要求

在命令行完成计算服务的基本操作任务，如正常启动、关闭和重建虚拟机等操作，在 Dashboard 界面上完成启动和部署云主机的任务。

核心概念

计算服务：计算服务组件 Nova 是 OpenStack 云中的计算组织控制器。支持

OpenStack 云中实例（Instances）生命周期的所有活动都由 Nova 处理。这样使得 Nova 成为一个负责管理计算资源、网络、认证、所需可扩展性的平台。

视频

计算服务的相关知识

知识准备

1. 概述

计算服务组件 Nova 可以说是整个云平台最重要的组件，基本功能包含运行虚拟机实例、管理网络以及通过用户和项目来控制对云的访问。

（1）功能如下：

①实例生命周期管理。

②管理计算资源。

③网络和认证管理。

④REST 风格的 API 管理。

⑤异步的一致性通信管理。

（2）特点。

①Hypervisor 透明：支持 Xen、XenServer/XCP、KVM、UML、VMware vSphere and Hyper－V。

②Nova 是 OpenStack 中最核心的组件。OpenStack 的其他组件归根结底是为 Nova 组件服务的。

③Nova 服务由多个子服务构成，子服务通过 RPC 实现通信。服务之间有很松的耦合性。

2. 计算服务架构

计算服务 Nova 的服务架构如图 6－1 所示。

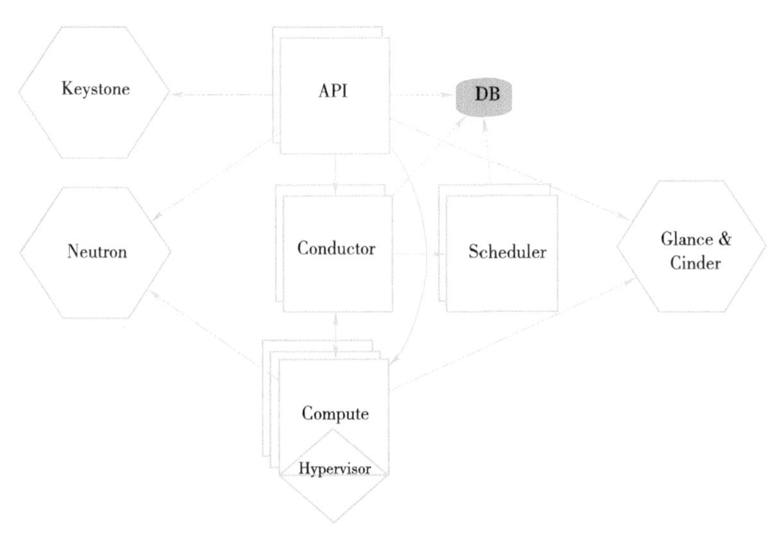

图 6－1　计算服务组件 Nova 的服务架构

（1）Nova API：HTTP 服务，用于接收和处理客户端发送的 HTTP 请求。

（2）Nova Compute：Nova 组件最核心的服务，实现虚拟机管理的功能。实现了在计算节点上创建、启动、暂停、关闭和删除虚拟机，虚拟机在不同的计算节点间迁移、虚拟机安全控制、管理虚拟机磁盘镜像以及快照等功能。

（3）Nova Cert：用于管理证书，为了兼容 AWS。AWS 提供一整套的基础设施和应用程服务，使得绝大多数的应用程序在云上运行。

（4）Nova Conductor：RPC 服务，主要提供数据库查询功能。

（5）Nova Scheduler：Nova 调度子服务。当客户端向 Nova 服务器发起创建虚拟机请求时，决定将虚拟机创建在哪个节点上。

（6）Rabbit MQ Server：OpenStack 节点之间通过消息队列使用 AMQP（Advanced Message Queue Protocol）完成通信。

Nova 通过异步调用请求响应，使用回调函数在收到响应时触发。因为使用了异步通信，不会有用户长时间处于等待状态。

（7）Nova Console、Nova Consoleauth、Nova VNCProxy：Nova 控制台子服务。功能是实现客户端通过代理服务器远程访问虚拟机实例的控制界面。

（8）nova - volume 是创建、挂载、卸载持久化的磁盘虚拟机，运行机制类似 nova - computer。同样是接收消息队列中的执行指令并执行。Volume 相关职责包括：创建硬盘、删除硬盘、弹性计算硬盘、为虚拟机增加块设备存储。

3. 计算服务的工作流程

（1）计算服务的运行架构，如图 6 - 2 所示。

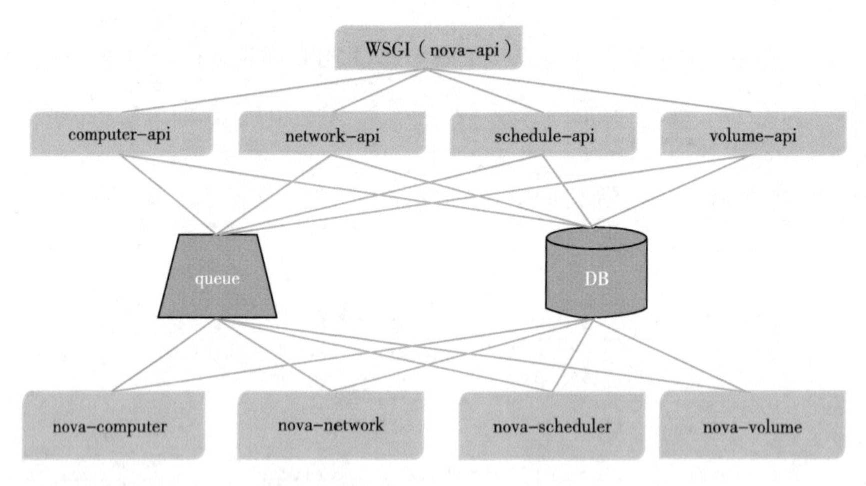

图 6 - 2　计算服务的运行架构

①nova - api 对外统一提供标准化接口、各子模块，如计算资源（Computer），存储资源（Volume）、网络资源（Network）模块通过相应的 API 接口服务对外提供服务。

②API 接口操作 DB 实现资源数据模型的维护。

③通过消息中间件，通知相应的守护进程如 nova - compute 实现服务接口，API 与守护进程共享 DB 数据库，但守护进程侧重维护状态信息，网络资源状态。

守护进程之间不能直接调用，需要通过 API 调用，如 nova - compute 为虚拟机分配网络，需要调用 network - api，而不能直接调用 nova - network，这样易于解耦合。

（2）以创建虚拟机为例：

①调用 nova－api 创建虚拟机接口，nova－api 对参数进行解析以及初步合法性校验。

②调用 compute－api 创建虚拟机 VM 接口，computer－api 根据虚拟机参数（CPU、内存、磁盘、网络、安全组等）信息，访问数据库创建数据模型虚拟机实例记录。

③computer－api 通过 RPC 的方式将创建虚拟机的基础信息封装成消息发送至消息中间件指定消息队列“scheduler”。

④nova－scheduler 订阅了消息队列“scheduler”的内容，接收到创建虚拟机的消息后，进行过滤，根据请求的虚拟资源，即 flavor 的信息，选择一台物理主机部署，如 novan1。nova－scheduler 将虚拟机的基本信息、所属物理主机信息发送至消息中间件指定消息队列“computer. novan1”。

⑤novan1 上 nova－compute 守护进程订阅消息队列“computer. novan1”，接收到消息后，根据虚拟机基本信息开始创建虚拟机。

⑥nova－computer 调用 network－api 分配网络 IP。

⑦nova－network 接收消息后，从 fixedIP 表中拿出一个可用的 IP，nova－network 根据私网资源池，结合 DHCP，实现 IP 分配和 IP 绑定。

⑧nova－computer 通过调用 volume－api 实现存储划分，最后调用底层虚拟化技术，部署虚拟机。

视频

计算服务的基本操作

1．实施任务的准备环境（见表 6－1）

表 6－1　实施任务的准备环境

实施任务所需软件资源	虚拟机镜像资源信息
虚拟机软件（vm12）	controller
securcrt 远程连接软件	compute

2．实施任务的简明过程（见表 6－2）

表 6－2　实施任务的简明过程

序号	步骤	详细操作及说明
1	计算服务的基本操作	①检测 Nova 服务是否安装； ②检测 Nova 服务列表； ③检测 Nova 服务的运行状态； ④Nova 管理镜像； ⑤Nova 管理安全组规则； ⑥Nova 管理虚拟机类型； ⑦Nova 对逻辑卷的管理
2	Dashboard 界面上启动和部署云主机，并测试	①上传镜像； ②配置一个网络架构

3. 计算服务的基本操作

(1) 检测 Nova 服务是否安装

```
[root@ controller ~] # source /etc/keystone/admin-openrc. sh
[root@ controller ~] # rpm -qa |grep nova
```

结果如图 6-3 所示。

```
[root@xiandian ~]# rpm -qa |grep nova
openstack-nova-novncproxy-13.1.0-1.el7.noarch
openstack-nova-common-13.1.0-1.el7.noarch
openstack-nova-console-13.1.0-1.el7.noarch
python-nova-13.1.0-1.el7.noarch
openstack-nova-api-13.1.0-1.el7.noarch
python-novaclient-3.3.1-1.el7.noarch
openstack-nova-compute-13.1.0-1.el7.noarch
openstack-nova-scheduler-13.1.0-1.el7.noarch
openstack-nova-conductor-13.1.0-1.el7.noarch
```

图 6-3 检测情况

(2) 检测 Nova 服务列表。

```
[root@ controller ~] # openstack-service list |grep nova
```

结果如图 6-4 所示。

```
[root@xiandian ~]# openstack-service list |grep nova
openstack-nova-api
openstack-nova-compute
openstack-nova-conductor
openstack-nova-consoleauth
openstack-nova-novncproxy
openstack-nova-scheduler
```

图 6-4 服务列表

(3) 检测 Nova 服务的运行状态。

```
[root@ controller ~] # openstack-service status |grep nova
```

结果如图 6-5 所示。

```
[root@xiandian ~]# openstack-service status |grep nova
MainPID=942 Id=openstack-nova-api.service ActiveState=active
MainPID=1216 Id=openstack-nova-compute.service ActiveState=active
MainPID=933 Id=openstack-nova-conductor.service ActiveState=active
MainPID=941 Id=openstack-nova-consoleauth.service ActiveState=active
MainPID=954 Id=openstack-nova-novncproxy.service ActiveState=active
MainPID=937 Id=openstack-nova-scheduler.service ActiveState=active
```

图 6-5 服务状态

(4) Nova 管理镜像。

①Nova 获取镜像列表。

```
[root@ controller ~] # nova image-list
```

结果如图 6-6 所示。

```
[root@xiandian ~]# nova image-list
+--------------------------------------+---------+--------+--------+
| ID                                   | Name    | Status | Server |
+--------------------------------------+---------+--------+--------+
| ea973ca4-8903-4bbb-9806-91997f65eea1 | CentOS7 | ACTIVE |        |
+--------------------------------------+---------+--------+--------+
```

图 6-6 获取镜像列表

②Nova 查询镜像详细信息。

镜像 ID 通过 nova image – list 查询得出。

```
[root@ controller ~] # nova image – show ea973ca4 – 8903 –
4bbb –9806 –91997f65eea1
```

结果如图 6 –7 所示。

```
[root@xiandian ~]# nova image-show ea973ca4-8903-4bbb-9806-91997f65eea1
+----------------------+--------------------------------------+
| Property             | Value                                |
+----------------------+--------------------------------------+
| OS-EXT-IMG-SIZE:size | 400752640                            |
| created              | 2019-05-29T05:51:37Z                 |
| id                   | ea973ca4-8903-4bbb-9806-91997f65eea1 |
| minDisk              | 0                                    |
| minRam               | 0                                    |
| name                 | CentOS7                              |
| progress             | 100                                  |
| status               | ACTIVE                               |
| updated              | 2019-05-29T05:51:39Z                 |
+----------------------+--------------------------------------+
```

图 6 –7　查询镜像详细信息

（5）Nova 管理安全组规则。

①创建安全组。

创建的语法格式：nova secgroup – create <name> <description>。

参数说明：

<name>：安全组名字；<description>：安全组描述。

```
[root@ controller ~] #nova secgroup –create nova –test 'test
the nova command about the rules'
```

结果如图 6 –8 所示。

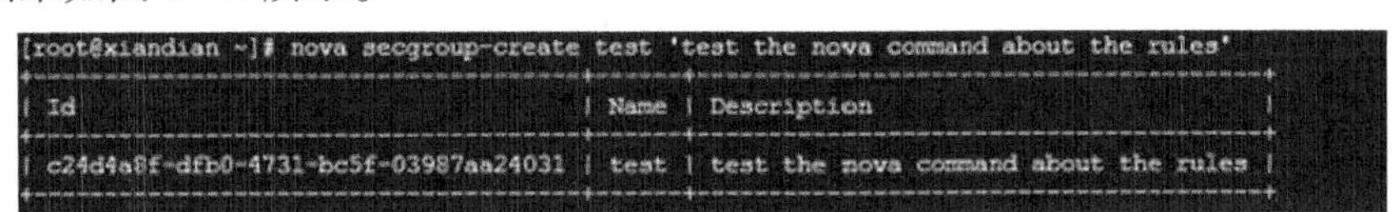

```
[root@xiandian ~]# nova secgroup-create test 'test the nova command about the rules'
+--------------------------------------+------+--------------------------------------+
| Id                                   | Name | Description                          |
+--------------------------------------+------+--------------------------------------+
| c24d4a8f-dfb0-4731-bc5f-03987aa24031 | test | test the nova command about the rules |
+--------------------------------------+------+--------------------------------------+
```

图 6 –8　创建安全组

②安全组添加规则。

添加语法格式：nova secgroup – add – rule <secgroup> <ip – proto> <from – port> <to – port> <cidr>

参数说明：

<secgroup>：安全组名字或者 ID；<ip – proto> IP：协议（icmp、tcp、udp）；<from – port>：起始端口；<to – port>：结束端口；<cidr>：网络地址。

```
[root@ controller ~] # nova secgroup – add – rule nova – test
icmp -1 -1 0. 0. 0. 0/0
```

结果如图 6 –9 所示。

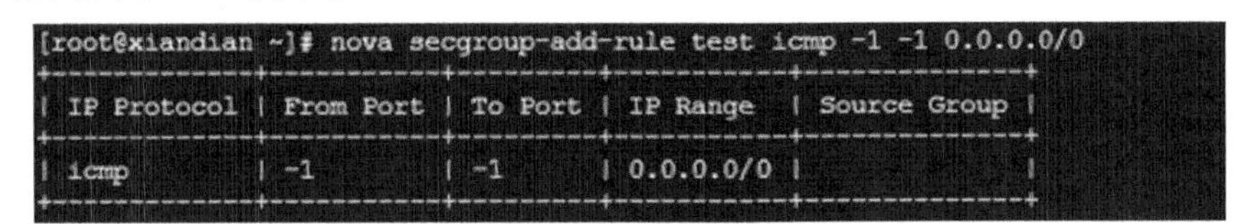

```
[root@xiandian ~]# nova secgroup-add-rule test icmp -1 -1 0.0.0.0/0
+-------------+-----------+---------+-----------+--------------+
| IP Protocol | From Port | To Port | IP Range  | Source Group |
+-------------+-----------+---------+-----------+--------------+
| icmp        | -1        | -1      | 0.0.0.0/0 |              |
+-------------+-----------+---------+-----------+--------------+
```

图 6 –9　添加规则

视频
创建和启动云主机

4. 在 Dashboard 界面上启动和部署云主机，并进行测试

（1）上传一个镜像。本地上传网页界面，下载网络地址，上传后台命令，如图 6－10 所示。

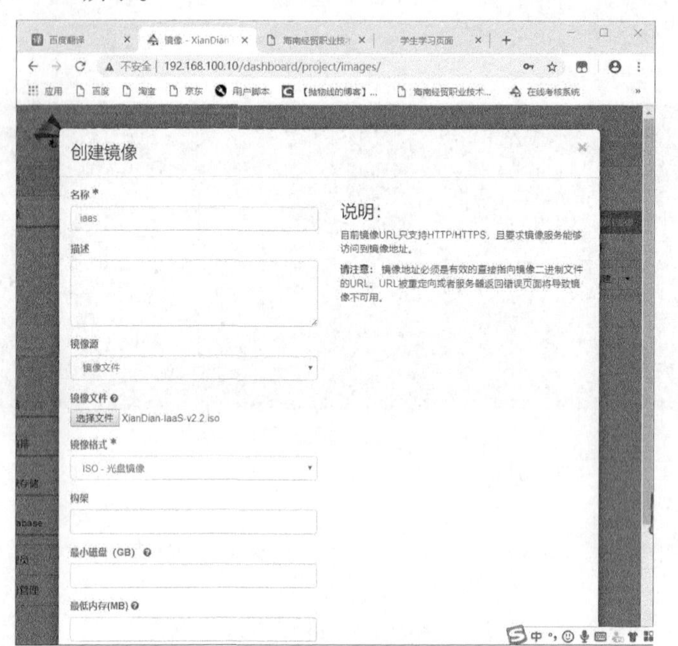

图 6－10　上传一个镜像

（2）配置一个网络架构。

①增加二层的外部网络。

首先，在 Dashboard 界面的“系统”菜单创建外部网络 ext－net，项目为 admin，供应商网络类型为 GRE，段 ID 为 1，并设置为共享和外部网络；其次，在“网络”菜单创建子网名为 ext－subnet，网络地址为 192.168.200/24，网关地址是 192.168.200.1，子网的范围为 192.168.200.100～192.168.200.200，如图 6－11～图 6－13 所示。

图 6－11　创建外部网络 1

网络
网络拓扑
网络
路由
负载均衡
防火墙
管理员
身份管理

名称
ID
项目ID
状态
管理状态
共享的
外部网络
MTU
供应商网络

子网
名称

端口
名称

创建子网

子网　子网详情

子网名称

ext-subnet

网络地址

192.168.200.0/24

IP版本

IPv4

网关IP

192.168.200.1

禁用网关

创建关联到这个网络的级配置。

« 返回

图 6－12　创建外部网络 2

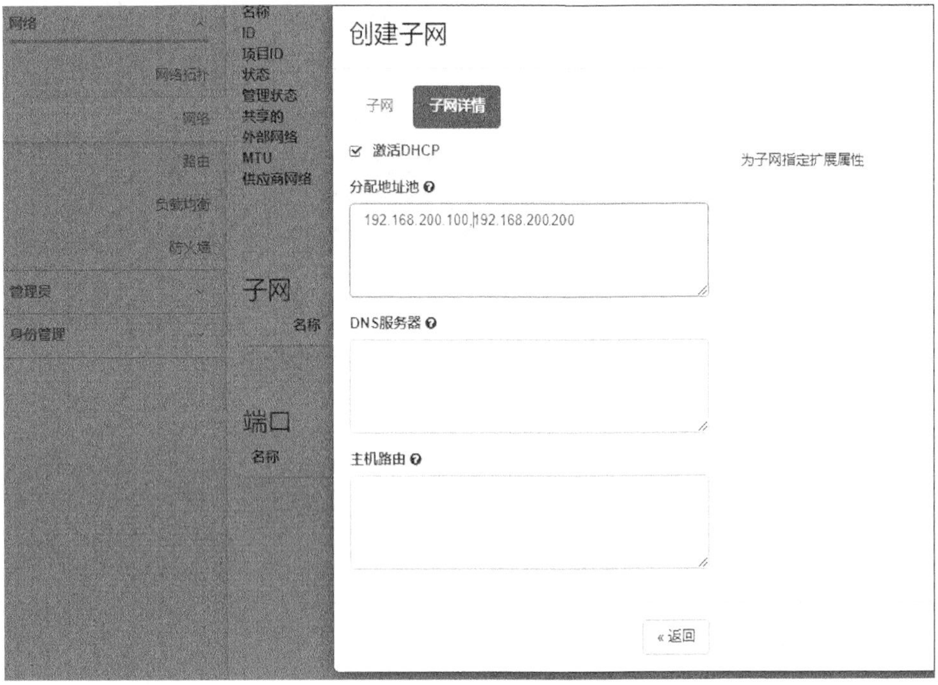

网络
网络拓扑
网络
路由
负载均衡
防火墙
管理员
身份管理

名称
ID
项目ID
状态
管理状态
共享的
外部网络
MTU
供应商网络

子网
名称

端口
名称

创建子网

子网　子网详情

激活DHCP

分配地址池

192.168.200.100,192.168.200.200

DNS服务器

主机路由

为子网指定扩展属性

« 返回

图 6－13　创建外部网络 3

②增加二层的内部网络。

首先，在 Dashboard 界面的“网络”菜单创建内部网络 int－net，其次在“子网”菜单创建子网名为 ext－subnet，网络地址为 10.0.0.0/24，网关地址为 10.0.0.1，其子网的范围为 10.0.0.100～10.0.0.200，如图 6－14～图 6－16 所示。

图 6－14　创建内部网络 1

图 6－15　创建内部网络 2

图 6-16　创建内部网络 3

③增加路由器 router，如图 6-17 所示。

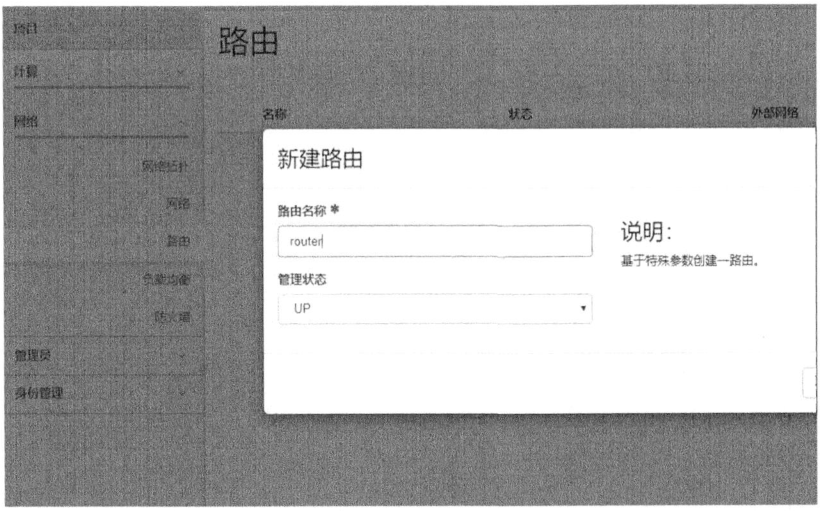

图 6-17　增加路由器 router

（3）部署和启动云主机，并进行测试。

首先，在 Dashboard 界面的“云主机”菜单创建云主机，名称为 test，源为 CentOS 7.2，favor 为 m1. small，网络为 int - net，安全组选择 default，然后选择下一步，最后启动实例，经过孵化 5 分钟左右（不同计算机的配置时间略有不同），当状态显示“运行”时，表示已经成功启动云主机，如图 6 - 18 ~ 图 6 - 23 所示。

图 6 - 18　部署和启动云主机 1

图 6 - 19　部署和启动云主机 2

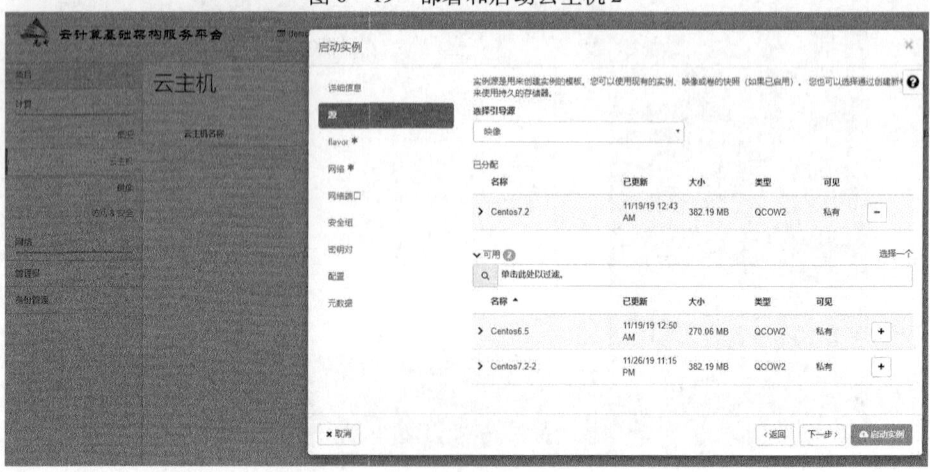

图 6 - 20　部署和启动云主机 3

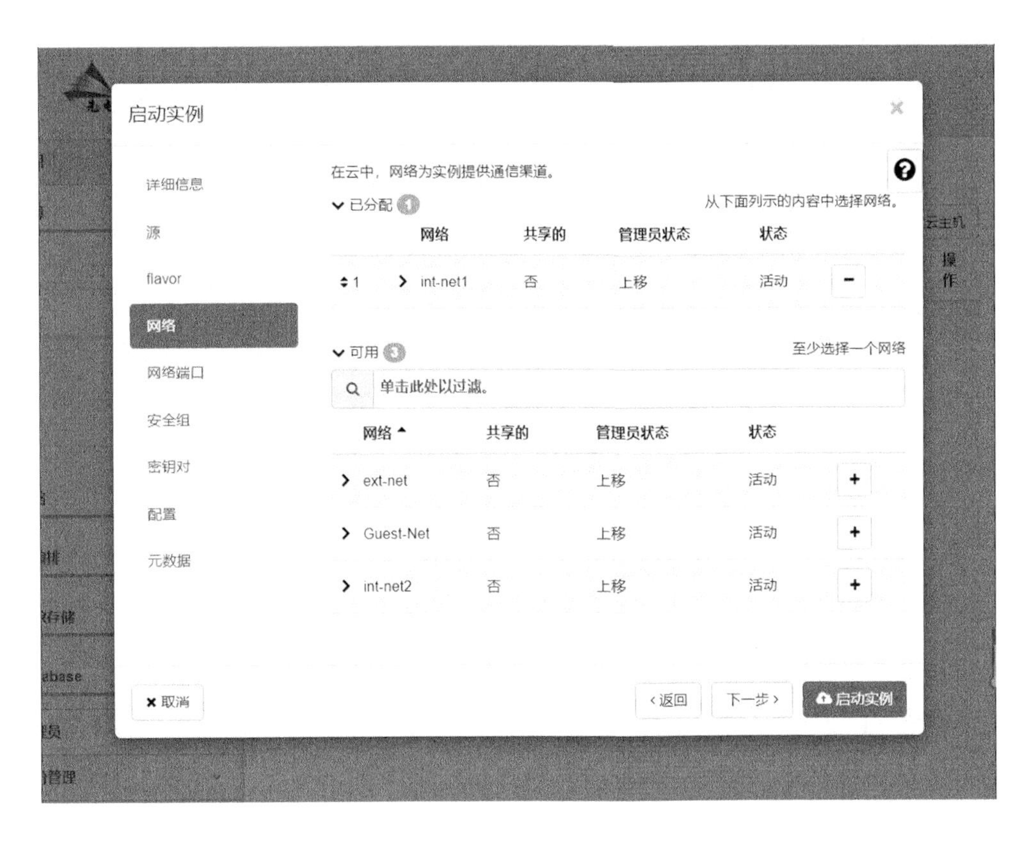

图 6－21　部署和启动云主机 4

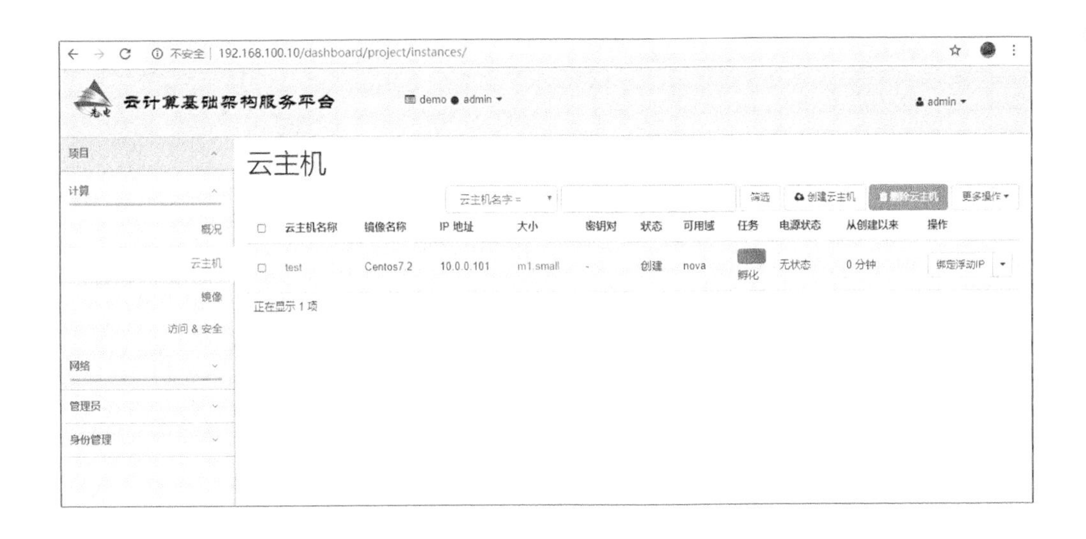

图 6－22　部署和启动云主机 5

图 6－23　部署和启动云主机 6

（4）测试网络部分。

测试网络是指让物理机可以通过外网 192.168.200 网段连接云主机。

①在“网络”菜单的“网络拓扑”的“路由器”选项下选择“增加接口”，使云主机和外网连接，如图 6－24 和 6－25 所示。

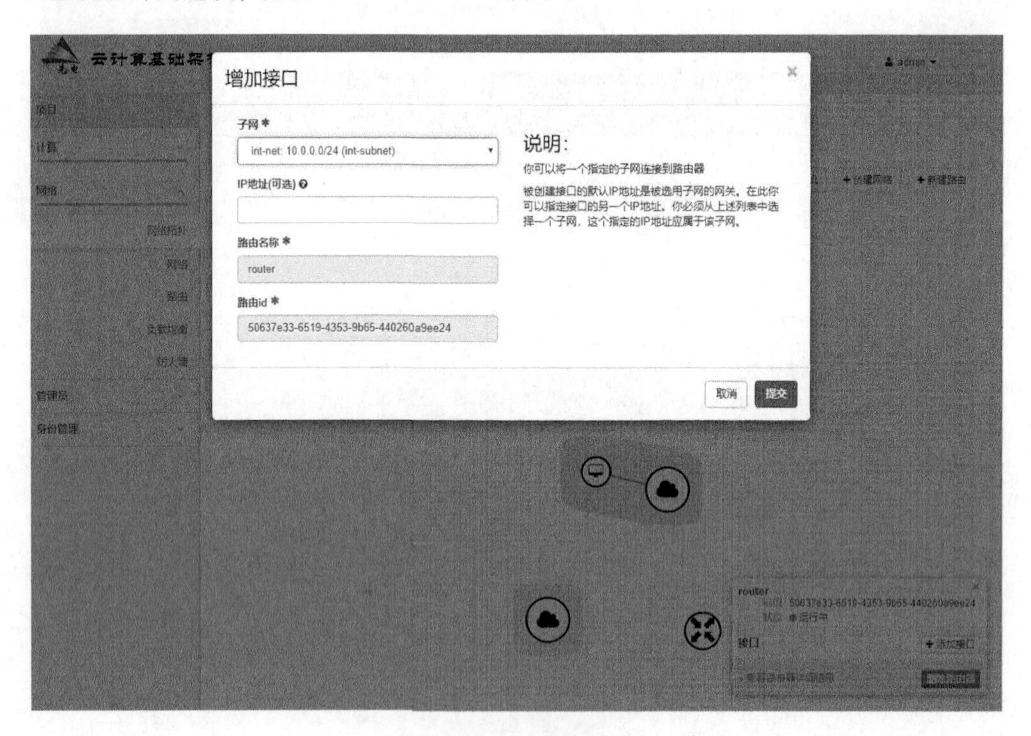

图 6－24　测试网络 1

②为云主机绑定浮动 IP，即外网的 IP 能连接云主机，如图 6－26 ~ 图 6－28 所示。

③设置终端软件能访问云主机。首先设置云主机的安全，然后添加上 ICMP、TCP、UDP 及 SSH 协议的出口与入口，允许这些协议能正常访问云主机，并且设置物理机的虚拟网卡 VMnet1 的 IP 为 192.168.200.102 的网段，最后使用终端软件，即可正常访问云主机，如图 6－29 ~ 图 6－31 所示。

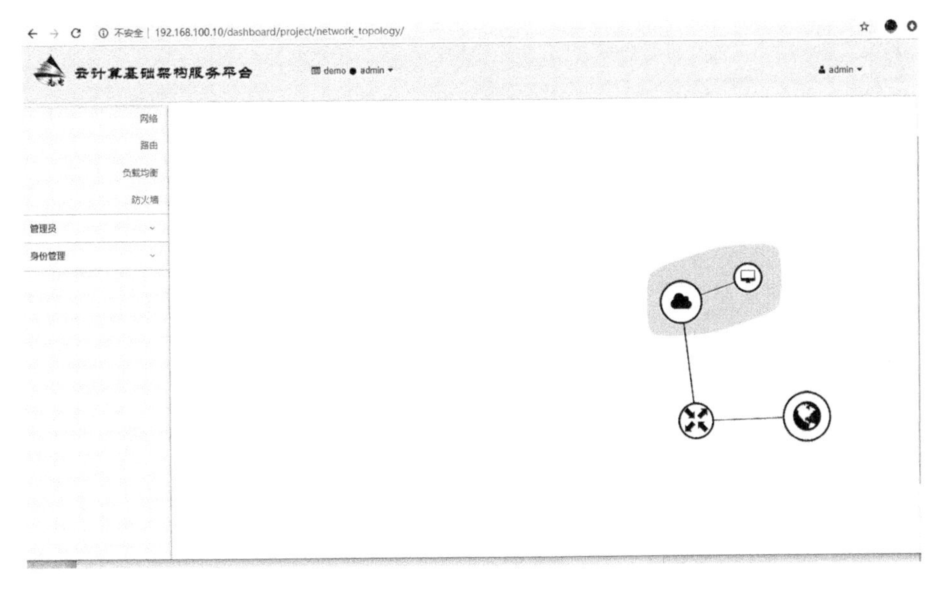

图 6－25　测试网络 2

图 6－26　云主机绑定浮动 IP1

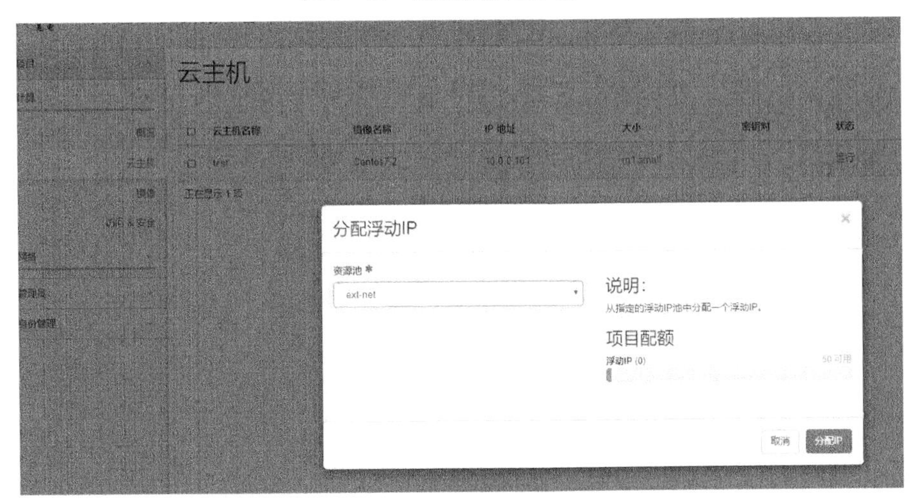

图 6－27　云主机绑定浮动 IP2

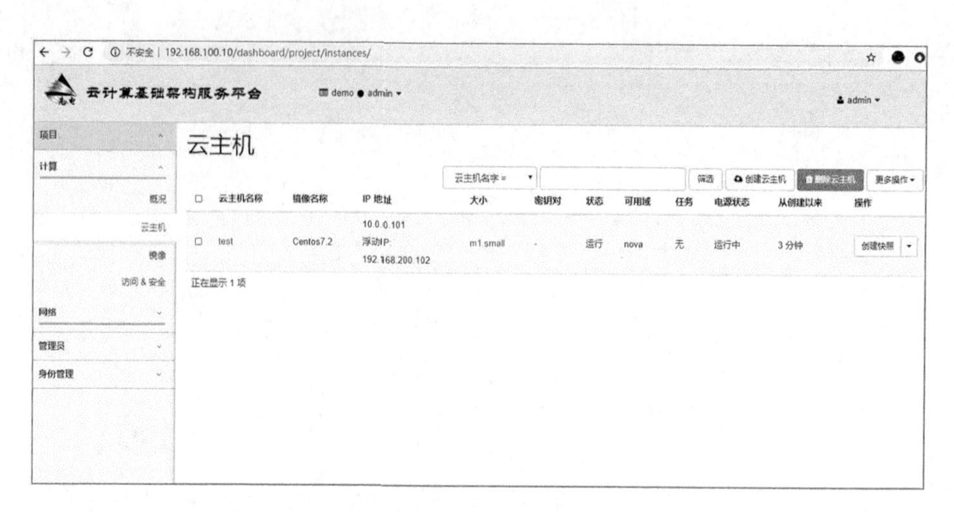

图 6－28　云主机绑定浮动 IP3

方向	以太网类型（EtherType）	IP协议	端口范围	远端IP前缀	远端安全组	操作
入口	IPv4	任何	任何	-	default	删除规则
出口	IPv6	任何	任何	::/0	-	删除规则
入口	IPv6	任何	任何	-	default	删除规则
出口	IPv4	任何	任何	0.0.0.0/0	-	删除规则
出口	IPv4	ICMP	任何	0.0.0.0/0	-	删除规则
入口	IPv4	ICMP	任何	0.0.0.0/0	-	删除规则
入口	IPv4	TCP	1 - 65535	0.0.0.0/0	-	删除规则
出口	IPv4	TCP	1 - 65535	0.0.0.0/0	-	删除规则
入口	IPv4	TCP	22 (SSH)	0.0.0.0/0	-	删除规则
入口	IPv4	UDP	1 - 65535	0.0.0.0/0	-	删除规则
出口	IPv4	UDP	1 - 65535	0.0.0.0/0	-	删除规则

正在显示 11 项

图 6－29　设置终端软件访问云主机 1

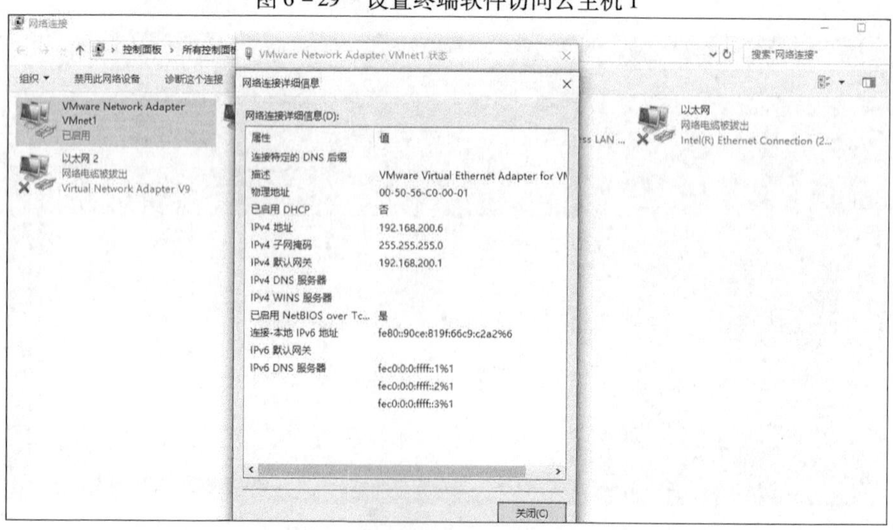

图 6－30　设置终端软件访问云主机 2

```
Last login: Fri Nov 29 11:42:18 2019

 #########################
 #  Welcome to XianDian  #
 #########################

[root@test ~]# ip a
1: lo: <LOOPBACK,UP,LOWER_UP> mtu 65536 qdisc noqueue state UNKNOWN
    link/loopback 00:00:00:00:00:00 brd 00:00:00:00:00:00
    inet 127.0.0.1/8 scope host lo
       valid_lft forever preferred_lft forever
    inet6 ::1/128 scope host
       valid_lft forever preferred_lft forever
2: eth0: <BROADCAST,MULTICAST,UP,LOWER_UP> mtu 1458 qdisc pfifo_fast st
ate UP qlen 1000
    link/ether fa:16:3e:81:aa:dc brd ff:ff:ff:ff:ff:ff
    inet 10.0.0.101/24 brd 10.0.0.255 scope global dynamic eth0
       valid_lft 85197sec preferred_lft 85197sec
    inet6 fe80::f816:3eff:fe81:aadc/64 scope link
       valid_lft forever preferred_lft forever
[root@test ~]#
```

图 6－31　设置终端软件访问云主机 3

项目小结

计算服务组件 Nova 是 OpenStack 最重要的项目，处于 OpenStack 的中心，本项目学习 OpenStack Nova 的功能和架构，讨论计算服务的工作流程，并通过任务实施使读者完成项目目标。

项目扩展

表 6－3 列举本项目 Nova 的常用命令及其说明。

表 6－3　常用命令及其说明

常用命令	命令说明
nova list	查看虚拟机
nova stop［vm－name］或［vm－id］	关闭虚拟机
nova start［vm－name］或［vm－id］	启动虚拟机
nova suspend［vm－name］或［vm－id］	暂停虚拟机
nova resume［vm－name］或［vm－id］	启动暂停的虚拟机
nova delete［vm－name］或［vm－id］	删除虚拟机
nova migrate	冷迁移
nova flavor－create	生成新的模板
nova flavor－delete	删除指定的模板
nova flavor－list	列举所有可用的模板
nova hypervisor－list	列举出所有计算节点的信息
nova hypervisor－servers	列举出一个计算节点的所有虚拟机
nova hypervisor－show	显示一台计算节点的详细信息
hypervisor－stats	获取所有计算节点的统计信息

项目思考

Nova 组件创建虚拟机的过程历经了哪些过程？

项目七 存储服务

项目综述

云的3种基本服务是计算服务、网络服务和存储服务，小张已经了解前两个了，需继续了解最后一个存储服务。在OpenStack中，有3个与存储相关的组件，Glance提供虚拟机镜像存储和管理，Cinder提供块存储，Swift提供对象存储。Glance已经介绍，本项目将介绍其他OpenStack的存储服务，同时了解内部机制和流程，使用存储为平台的其他服务提供后端存储和数据的安全备份。小张为此要完成的任务如下：

视频

项目七综述

- 利用Cinder为虚拟机提供虚拟磁盘。
- 利用Swift为虚拟机提供虚拟磁盘。

项目目标

【知识目标】

- OpenStack的Cinder和Swift服务组件的相关概念。
- OpenStack的Cinder云硬盘和Swift对象存储服务流程和工作机制。

【技能目标】

- Cinder服务的基本操作。
- Swift服务的基本操作。
- 在命令行CLI完成块存储和对象存储任务。

【职业能力】

根据企业的应用需求，利用Cinder为虚拟机提供硬盘空间。

任务一　安装与操作块服务

任务要求

块存储是将磁盘整个映射给主机使用，提供逻辑卷功能。本功能主要针对OpenStack的块存储服务进行剖析。

核心概念

块服务：OpenStack 块服务 Cinder 的核心功能是给云主机增加云硬盘的服务，它同时也对卷进行管理，允许对卷、卷的类型、快照进行处理。但它并没有实现对块设备的管理和实际服务，而是通过后端的统一存储接口支持不同块设备厂商的块存储服务，实现其驱动支持并与 OpenStack 进行整合。

知识准备

块存储的相关知识

1. 基本概念

Cinder 是 OpenStack 中提供块存储服务的组件，主要是为虚拟机实例提供虚拟磁盘。

在 OpenStack 中提供对卷从创建到删除整个生命周期的管理，从虚拟机实例的角度来看，挂载的每一个卷都是一块硬盘。OpenStack 提供块存储服务的是 Cinder，具体功能如下：

（1）提供 API 使用户能够查询和管理卷、卷快照以及卷类型。

（2）提供 scheduler 调度卷创建请求，合理优化存储资源的分配。

（3）通过 driver 架构支持多种后端存储方式。

2. 架构讲解

Cinder 的架构如图 7－1 所示。

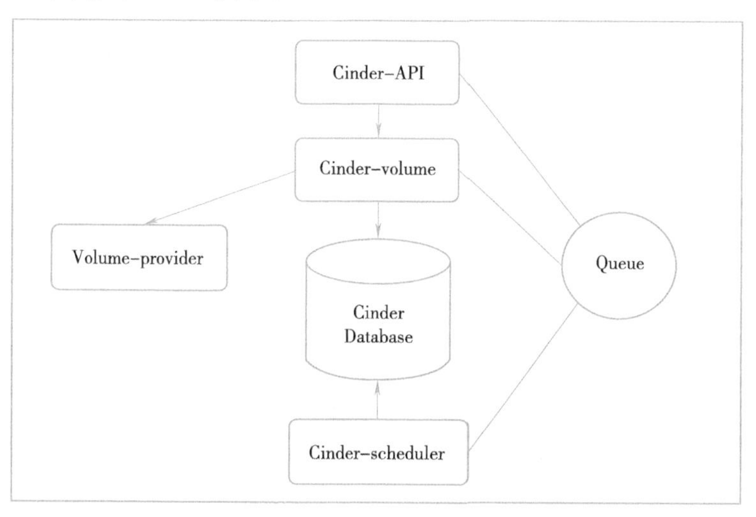

图 7－1　Cinder 架构

Cinder 包含如下几个组件：

（1）Cinder－API：接收 API 请求，调用 Cinder－volume 执行操作。

（2）Cinder－volume：管理 volume 的服务，与 Volume－provider 协调工作，管理 volume 的生命周期。运行 Cinder－volume 服务的节点被称为存储节点。

（3）Cinder－scheduler：scheduler 通过调度算法选择最合适的存储节点创建 volume。

（4）Volume－provider：数据的存储设备，为 volume 提供物理存储空间。Cinder－volume 支持多种 Volume－provider，每种 Volume－provider 通过自己的 driver 与 Cinder－

volume 协调工作。

（5）Queue：Cinder 各个子服务通过消息队列实现进程间通信和相互协作。因为有了消息队列，子服务之间实现了解耦，这种松散的结构也是分布式系统的重要特征。

（6）Cinder Database：Cinder 有一些数据需要存放到数据库中，一般使用 MySQL。数据库是安装在控制节点上的，比如在实验环境中，可以访问名称为 cinder 的数据库。

3. Cinder 流程

Cinder 流程如图 7－2 所示。

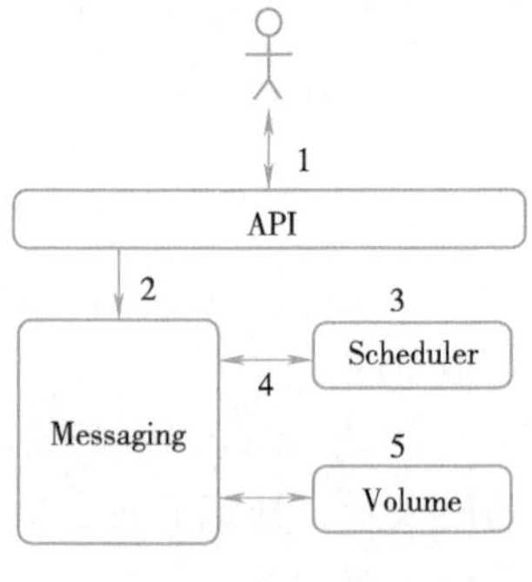

图 7－2　Cinder 流程

步骤 1：客户（可以是 OpenStack 最终用户，也可以是其他程序）向 API（Cinder－API）发送请求，帮用户创建一个 volume。

步骤 2：API 对请求做一些必要处理后，向 Messaging（RabbitMQ）发送一条消息，让 Scheduler 创建一个 volume。

步骤 3：Scheduler（Cinder－scheduler）从 Messaging（RabbitMQ）获取到 API 发送的消息，然后执行调度算法，从若干计算节点中选出节点 A。

步骤 4：Scheduler 向 Messaging 发送一条消息，让存储节点 A 创建这个 volume。

步骤 5：存储节点 A 的 Volume（Cinder－volume）从 Messaging 中获取 Scheduler 发送的消息，然后通过 driver 在 Volume－provider 上创建 volume。

4. Cinder 组件详细介绍

（1）Cinder－API。Cinder－API 是整个 cinder 组件的门户，所有的 Cinder 的请求都首先由 nova－API 处理，它向外界暴露了若干 HTTP REST API 接口，在 Keystone 中可以查询到 Cinder－API 的服务端点，如图 7－3 所示。

```
[root@localhost ~(keystone_admin)]# openstack endpoint list --service cinder
+----------------------------------+-----------+--------------+--------------+---------+-----------+----
----------------------------------------+
| ID                               | Region    | Service Name | Service Type | Enabled | Interface | URL
                                        |
+----------------------------------+-----------+--------------+--------------+---------+-----------+----
----------------------------------------+
| 31dfd8bd50c44fa8bbaf173bbd49d0a5 | RegionOne | cinder       | volume       | True    | admin     | htt
p://172.31.208.100:8776/v1/%(tenant_id)s |
| 8bb4136e1fe34a4eb2c89f95f4f8cf23 | RegionOne | cinder       | volume       | True    | public    | htt
p://172.31.208.100:8776/v1/%(tenant_id)s |
| baf4d5f0cdcf4cecaedb684408204156 | RegionOne | cinder       | volume       | True    | internal  | htt
p://172.31.208.100:8776/v1/%(tenant_id)s |
```

图 7－3　Cinder－API 服务端点

客户端可以将请求发送到服务端点指定的地址，向 Cinder－API 请求操作，大部分的 API 请求都可以在 Dashboard 上进行，Cinder－API 对接收到的 HTTP API 请求会做如下处理：

①检查客户端传入的参数是否合法有效。

②调用 cinder 其他子服务的处理客户端请求。

③将 cinder 其他子服务返回的结果序列号返回给客户端。

（2）cinder－scheduler。创建 volume 时，cinder－scheduler 会基于容量、volume type 等条件选择出最合适的存储节点，然后让其创建 volume。

（3）cinder - volume。OpenStack 中对 volume 的操作，最后都是交给 cinder - volume 来完成的。

cinder - volume 自身并不管理整个的存储设备，存储设备是由 volume provider 管理的，cinder - volume 与 volume provider 一起实现 volume 生命周期管理，它通过 driver 架构支持多种 volume provider。

cinder - volume 为这些 volume provider 定义了统一的接口，volume provider 只需要实现这些接口，就可以 Driver 的形式即插即用到 OpenStack 系统中，如图 7 - 4 所示。

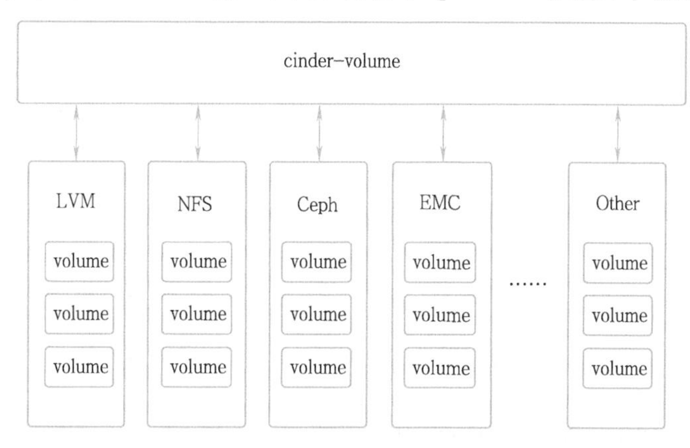

图 7 - 4　cinder - volume 的接口

5. LVM 技术

逻辑卷管理器（LVM 技术）是 Linux 系统用于对硬盘分区进行管理的一种机制，理论性较强，其创建初衷是为了解决硬盘设备在创建分区后不易修改分区大小的缺陷。尽管对传统的硬盘分区进行强制扩容或缩容从理论上来讲是可行的，但是却可能造成数据的丢失。而 LVM 技术是在硬盘分区和文件系统之间添加了一个逻辑层，它提供了一个抽象的卷组，可以把多块硬盘进行卷组合并。这样一来，用户不必关心物理硬盘设备的底层架构和布局，就可以实现对硬盘分区的动态调整。LVM 的技术架构如图 7 - 5 所示。

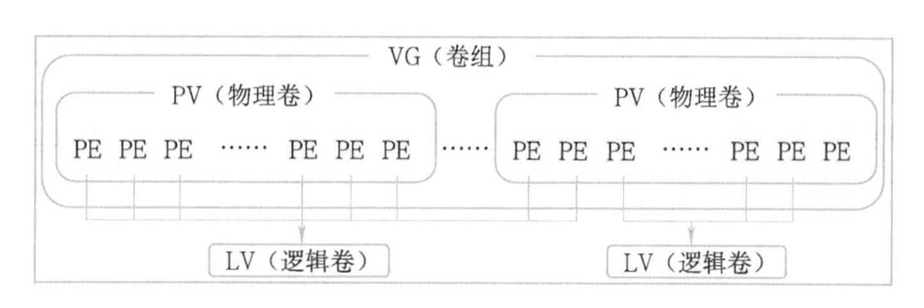

图 7 - 5　LVM 的技术架构

为了理解 LVM，举例说明。比如小明想吃馒头但是家里面粉不够了，于是妈妈从隔壁老王家、老李家、老张家分别借来一些面粉，准备蒸馒头吃。首先需要把这些面粉（物理卷［Physical Volume，PV］）揉成一个大面团（卷组［Volume Group，VG］），然后再把这个大面团分割成一个个小馒头（逻辑卷［Logical Volume，LV］），而且每个小馒头的质量必须是每勺面粉（基本单元［Physical Extent，PE］）的倍数。

物理卷处于 LVM 中的底层，可以将其理解为物理硬盘、硬盘分区或者 RAID 磁盘阵列，这都可以。卷组建立在物理卷之上，一个卷组可以包含多个物理卷，而且在卷组创建之后也可以继续向其中添加新的物理卷。逻辑卷是用卷组中空闲的资源建立的，并且逻辑卷在建立后可以动态地扩展或缩小空间。这就是 LVM 的核心理念。

6. Cinder 支持的后端存储类型

从实现来看，Cinder 对本地存储和 NAS 的支持比较不错，可以提供完整的 Cinder API V2 支持，而对于其他类型的存储设备，Cinder 的支持会或多或少受到限制。Rackspace 对于 Private Cloud 存储给出的典型配置如下：

（1）本地存储。

对于本地存储，cinder－volume 可以使用 lvm 驱动，该驱动当前的实现需要在主机上事先用 lvm 命令创建一个 cinder－volumes 的 vg，当该主机接收到创建卷请求时，cinder－volume 在该 vg 上创建一个 LV，并且用 openiscsi 将这个卷当作一个 iscsi tgt 给 export。当然还可以将若干主机的本地存储用 sheepdog 虚拟成一个共享存储，然后使用 sheepdog 驱动。

（2）其他存储。除了本地存储之外，还包括 EMC（EMC 是传统存储厂商，主要面对企业级用户）、Netapp（为数据密集型企业提供统一存储解决方案居世界前列的公司）和华为存储为之服务。

（3）Cinder Volume 创建流程。当用户从 cinder－client 发送一个请求创建卷组之后，会将创建的信息通过 REST 接口访问 Cinder－API，API 接受请求之后就会把 client 传送的请求进行解析，之后通过 RPC 将请求发送给 cinder－scheduler 选择合适的 volume 节点，节点选择完毕再次通过 RPC 发送给 volume，之后调用 driver 来创建 client 要求的卷组，如果 cinder 需要 backup，这时就需要调用 RPC 进行 backup 操作，如图 7－6 所示。

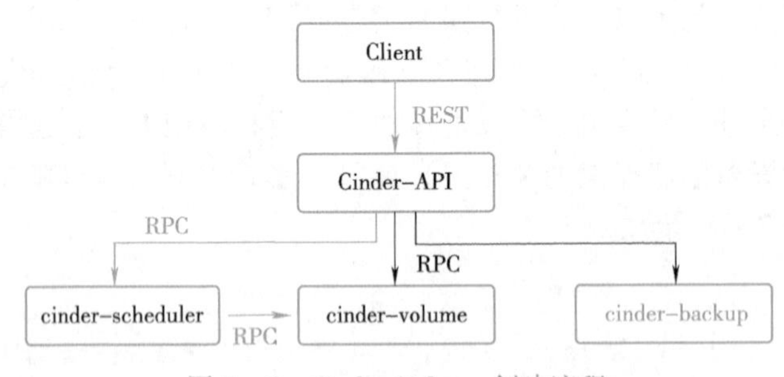

图 7－6　Cinder Volume 创建流程

1. 准备实施任务的环境（见表 7－1）

表 7－1　实施任务的准备环境

实施任务所需软件资源	虚拟机镜像资源信息
虚拟机软件（vm12）	controller
securcrt 远程连接软件	compute

2. 分解实施任务的过程（见表7-2）

表7-2　实施任务的简明过程

序　号	步　骤	详细操作及说明
1	块服务的基本操作	安装Cinder组件
2	命令行CLI完成块存储任务	①对Cinder后端逻辑卷进行扩容； ②扩容到50 GB； ③指定Cinder卷类型； ④创建并查看Cinder快照信息
3	在Dashboard界面上完成块存储任务	①对Cinder后端逻辑卷进行扩容； ②扩容到50 GB； ③指定Cinder卷类型； ④创建并查看Cinder快照信息

3. 块服务安装命令操作

安装Cinder组件。

①让环境变量生效。

```
[root@ controller ~] #source /etc/keystone/admin-openrc.sh
```

②挂载iso文件，如果/opt有镜像文件就无须挂载，如图7-7所示。

```
[root@ controller ~] #mount -o loop/opt/CentOS-7-x86_64-DVD-1511.iso /opt/centos7
[root@ controller ~] #mount  -o loop/opt/XianDian-IaaS-v2.2.iso   /opt/iaas
[root@ controller ~] #cd  /usr/local/bin
```

```
"local.repo" 10L, 168C written
[root@reboot opt]# mount -o loop /opt/CentOS-7-x86_64-DVD-1511.iso /opt/centos/
mount: /dev/loop0 is write-protected, mounting read-only
[root@reboot opt]# mount -o loop /opt/XianDian-IaaS-v2.2.iso /opt/iaas/
mount: /dev/loop1 is write-protected, mounting read-only
[root@reboot opt]#
```

图7-7　挂载iso文件

③在controller节点上修改openrc.sh文件，并安装Cinder组件，如图7-8和图7-9所示。

```
[root@ controller ~] #vi /etc/xiandian/openrc.sh
```

按【I】键进入编译模式，将文件修改为以下形式：

```
CINDER_DBPASS=000000
CINDER_PASS=000000
BLOCK_DISK=sdb
```

按【Esc】键并输入：wq命令保存退出。

在 controller 节点上执行 iaas - install - cinder - controller. sh，安装 Cinder 组件。

```
##--------------------Cinder Config--------------------##
##Password for Mysql cinder user. exmaple:000000
CINDER_DBPASS=000000

##Password for Keystore cinder user. exmaple:000000
CINDER_PASS=000000

##Cinder Block Disk. example:md126p3
BLOCK_DISK= sdb
```

图 7 - 8　修改 openrc. sh 文件

```
[root@controller bin]# ls
iaas-install-alarm.sh                    iaas-install-neutron-controller-flat.sh
iaas-install-ceilometer-compute.sh       iaas-install-neutron-controller-gre.sh
iaas-install-ceilometer-controller.sh    iaas-install-neutron-controller.sh
iaas-install-cinder-compute.sh           iaas-install-neutron-controller-vlan.sh
iaas-install-cinder-controller.sh        iaas-install-nova-compute.sh
iaas-install-dashboard.sh                iaas-install-nova-controller.sh
iaas-install-glance.sh                   iaas-install-swift-compute.sh
iaas-install-heat.sh                     iaas-install-swift-controller.sh
iaas-install-keystone.sh                 iaas-install-trove.sh
iaas-install-mysql.sh                    iaas-pre-host.sh
iaas-install-neutron-compute-flat.sh     iaas-uninstall-all.sh
iaas-install-neutron-compute-gre.sh      install.sh
iaas-install-neutron-compute.sh          server.yml
iaas-install-neutron-compute-vlan.sh
[root@controller bin]# install-Cinder-controller.sh
```


图 7 - 9　安装 Cinder 组件

④关闭 compute 虚拟机，并增加一个 20 GB 的硬盘，重启虚拟机，如图 7 - 10 和图 7 - 11 所示。

图 7 - 10　增加一个 20 GB 的硬盘

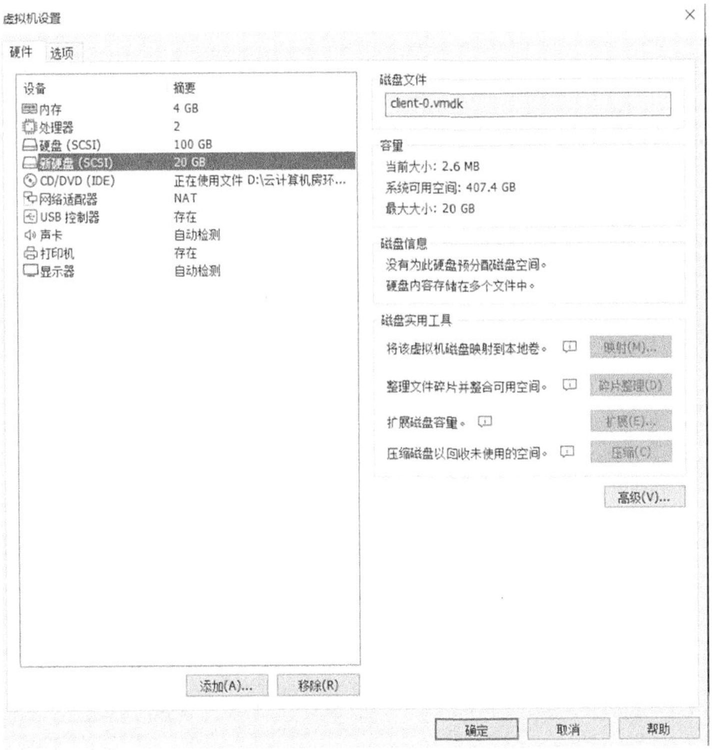

虚拟机设置

硬件 选项

设备	摘要
内存	4 GB
处理器	2
硬盘 (SCSI)	100 GB
新硬盘 (SCSI)	20 GB
CD/DVD (IDE)	正在使用文件 D:\云计算机房环...
网络适配器	NAT
USB 控制器	存在
声卡	自动检测
打印机	存在
显示器	自动检测

磁盘文件

client-0.vmdk

容量

当前大小：2.6 MB

系统可用空间：407.4 GB

最大大小：20 GB

磁盘信息

没有为此硬盘预分配磁盘空间。

硬盘内容存储在多个文件中。

磁盘实用工具

将该虚拟机磁盘映射到本地卷。 映射(M)...

整理文件碎片并整合可用空间。 碎片整理(D)

扩展磁盘容量。 扩展(E)...

压缩磁盘以回收未使用的空间。 压缩(C)

高级(V)...

添加(A)... 移除(R)

确定 取消 帮助

图 7－11 修改 openrc. sh 文件

在 computer 节点上配置分区，安装 Cinder 组件执行 IaaS － install － cinder － compute，同时更新与服务器 controller 节点的时间，使时间同步。最后重启计算节点，如图 7－12～图 7－14 所示。

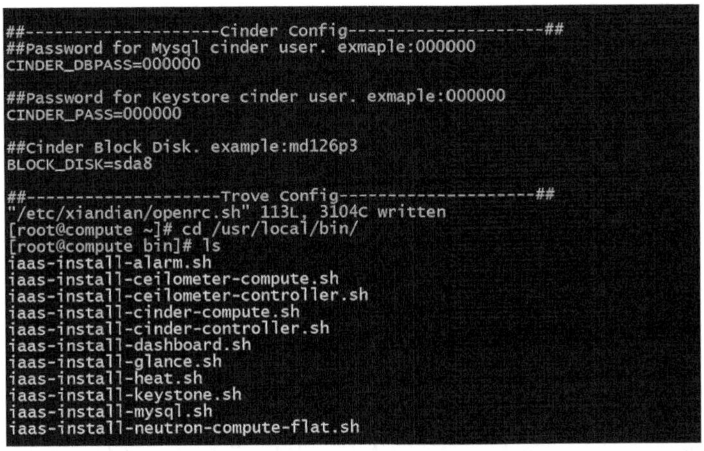

```
##--------------------Cinder Config--------------------##
##Password for Mysql cinder user. exmaple:000000
CINDER_DBPASS=000000

##Password for Keystore cinder user. exmaple:000000
CINDER_PASS=000000

##Cinder Block Disk. example:md126p3
BLOCK_DISK=sda8

##--------------------Trove Config--------------------##
"/etc/xiandian/openrc.sh" 113L, 3104C written
[root@compute ~]# cd /usr/local/bin/
[root@compute bin]# ls
iaas-install-alarm.sh
iaas-install-ceilometer-compute.sh
iaas-install-ceilometer-controller.sh
iaas-install-cinder-compute.sh
iaas-install-cinder-controller.sh
iaas-install-dashboard.sh
iaas-install-glance.sh
iaas-install-heat.sh
iaas-install-keystone.sh
iaas-install-mysql.sh
iaas-install-neutron-compute-flat.sh
```

图 7－12 安装 Cinder 组件

```
[root@ compute ~] #vi /etc/xiandian/openrc.sh
```

按【I】键进入编译模式，将文件修改为以下形式：

```
CINDER_ DBPASS =000000
CINDER_ PASS =000000
BLOCK_ DISK = sdb
```

按【Esc】键并输入：wq 命令保存退出。

```
[root@ compute ~] # cd /usr/local/bin
[root@ compute bin] # iaas - install - cinder - compute.sh
[root@ compute bin] # ntpdate controller
[root@ compute bin] # service openstack - cinder - volume restart
```

```
[root@compute bin]# ntpdate controller
26 Apr 22:28:59 ntpdate[110861]: no server suitable for synchronization
 found
[root@compute bin]#  service openstack-cinder-volume restart
Redirecting to /bin/systemctl restart  openstack-cinder-volume.service
[root@compute bin]#
```

图 7 – 13　同步 controller 服务器时间和重启计算节点

⑤在控制节点查看各节点状态是否 UP 状态，验证安装是否正确，如图 7 – 14 所示。

```
[root@ controller bin] # cinder service - list
```

```
[root@controller bin]# cinder service-list
+------------------+------------+------+---------+-------+----------------------------+-----------------+
|      Binary      |    Host    | Zone |  Status | State |         Updated_at         | Disabled Reason |
+------------------+------------+------+---------+-------+----------------------------+-----------------+
| cinder-scheduler | controller | nova | enabled |   up  | 2019-08-29T12:16:30.000000 |        -        |
|  cinder-volume   | compute@lvm| nova | enabled |   up  | 2019-08-29T12:16:26.000000 |        -        |
+------------------+------------+------+---------+-------+----------------------------+-----------------+
[root@controller bin]#
```

图 7 – 14　查看各节点的状态

4. 命令行 CLI 完成块存储任务

命令行CLI完成块存储任务

对 Cinder 后端逻辑卷进行扩容。

①创建一个 10 GB 的云硬盘并查看，如图 7 – 15 所示。

```
[root@controller bin]# cinder create --display-name cinder-volume-demo 10
+--------------------------------+--------------------------------------+
|            Property            |                Value                 |
+--------------------------------+--------------------------------------+
|          attachments           |                  []                  |
|       availability_zone        |                 nova                 |
|            bootable            |                false                 |
|      consistencygroup_id       |                 None                 |
|           created_at           |      2019-08-29T12:17:04.000000      |
|          description           |                 None                 |
|           encrypted            |                False                 |
|               id               | 6273c549-a274-419c-b021-4cb1521fc644 |
|            metadata            |                  {}                  |
|        migration_status        |                 None                 |
|          multiattach           |                False                 |
|              name              |          cinder-volume-demo          |
|     os-vol-host-attr:host      |                 None                 |
| os-vol-mig-status-attr:migstat |                 None                 |
| os-vol-mig-status-attr:name_id |                 None                 |
|  os-vol-tenant-attr:tenant_id  |   0fe13ac26d114f9fa91c29e02f09fbd1   |
|       replication_status       |               disabled               |
|              size              |                  10                  |
|          snapshot_id           |                 None                 |
|          source_volid          |                 None                 |
|             status             |               creating               |
|           updated_at           |                 None                 |
|            user_id             |   934a974a70d947949e8790d7442a4e37   |
|          volume_type           |                 None                 |
+--------------------------------+--------------------------------------+
[root@controller bin]# cinder list
+--------------------------------------+-----------+--------------------+------+-------------+----------+-------------+
|                  ID                  |   Status  |        Name        | Size | Volume Type | Bootable | Attached to |
+--------------------------------------+-----------+--------------------+------+-------------+----------+-------------+
| 6273c549-a274-419c-b021-4cb1521fc644 | available | cinder-volume-demo |  10  |      -      |  false   |             |
+--------------------------------------+-----------+--------------------+------+-------------+----------+-------------+
```

图 7 – 15　创建一个 10 GB 的云硬盘

```
[root@ controller bin] # cinder create - - display - name cinder
- volume - demo 10
[root@ controller bin] #cinder list
```

②扩容至 50 GB。

a. 计算节点新增一个 50 GB 硬盘，并重启 compute，如图 7 - 16 所示。

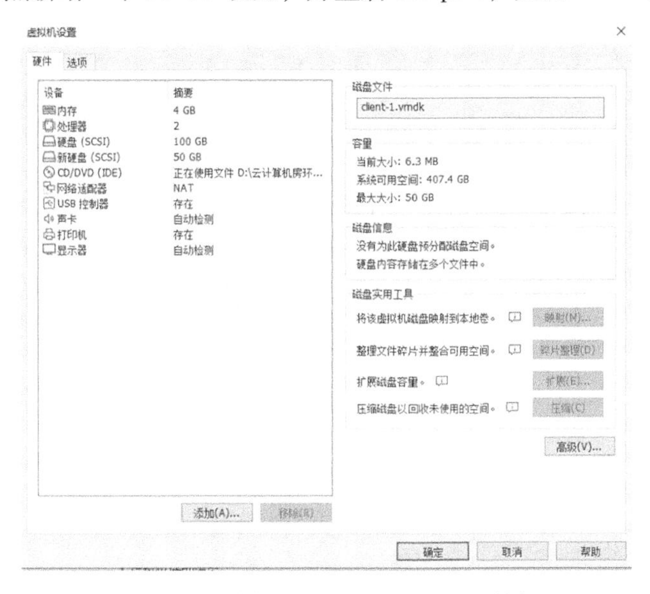

图 7 - 16　计算节点新增一个 50 GB 硬盘

b. 在 compute 节点，创建 LVM 物理卷和 cinder - volumes 卷组，输入如下命令，如图 7 - 17 所示。

```
[root@ compute ~] #pvcreate /dev/sdc
[root@ compute ~] #vgextend cinder -volumes /dev/sdc
[root@ compute ~] #vgdisplay
```

```
[root@compute ~]#
[root@compute ~]# pvcreate /dev/sdc
  Physical volume "/dev/sdc" successfully created
[root@compute ~]# vgextend cinder-volumes /dev/sdc
  Volume group "cinder-volumes" successfully extended
[root@compute ~]# Vgdisplay
-bash: Vgdisplay: command not found
[root@compute ~]# vgdisplay
  --- Volume group ---
  VG Name               cinder-volumes
  System ID
  Format                lvm2
  Metadata Areas        2
  Metadata Sequence No  3
  VG Access             read/write
  VG Status             resizable
  MAX LV                0
  Cur LV                1
  Open LV               0
  Max PV                0
  Cur PV                2
  Act PV                2
  VG Size               69.99 GiB
  PE Size               4.00 MiB
  Total PE              17918
  Alloc PE / Size       2560 / 10.00 GiB
  Free  PE / Size       15358 / 59.99 GiB
  VG UUID               6nBCP1-fORa-J8Jn-7Mi9-A0g3-oKrs-r5g6RU

  --- Volume group ---
  VG Name               centos
  System ID
  Format                lvm2
  Metadata Areas        1
  Metadata Sequence No  3
  VG Access             read/write
  VG Status             resizable
  MAX LV                0
  Cur LV                2
  Open LV               2
  Max PV                0
  Cur PV                1
```

图 7 - 17　创建 LVM 物理卷和 cinder - volumes 卷组

c. 在控制节点扩容至 50 GB，输入以下命令，如图 7－18 所示。

```
[root@ controller bin] # cinder create - - display - name cinder - volume - demo2 50
```

```
[root@controller bin]# cinder create --display-name cinder-volume-demo2 50
+--------------------------------+---------------------------------------+
| Property                       | Value                                 |
+--------------------------------+---------------------------------------+
| attachments                    | []                                    |
| availability_zone              | nova                                  |
| bootable                       | false                                 |
| consistencygroup_id            | None                                  |
| created_at                     | 2019-08-29T12:25:21.000000            |
| description                    | None                                  |
| encrypted                      | False                                 |
| id                             | bc605848-a9dd-4af7-ba66-3a2744b40103  |
| metadata                       | {}                                    |
| migration_status               | None                                  |
| multiattach                    | False                                 |
| name                           | cinder-volume-demo2                   |
| os-vol-host-attr:host          | compute@lvm#LVM                       |
| os-vol-mig-status-attr:migstat | None                                  |
| os-vol-mig-status-attr:name_id | None                                  |
| os-vol-tenant-attr:tenant_id   | 0fe13ac26d114f9fa91c29e02f09fbd1      |
| replication_status             | disabled                              |
| size                           | 50                                    |
| snapshot_id                    | None                                  |
| source_volid                   | None                                  |
| status                         | creating                              |
| updated_at                     | 2019-08-29T12:25:24.000000            |
| user_id                        | 934a974a70d947949e8790d7442a4e37      |
| volume_type                    | None                                  |
+--------------------------------+---------------------------------------+
[root@controller bin]#
```

图 7－18　在控制节点扩容至 50 GB

③指定 Cinder 卷类型。

a. 创建 lvm 标识的卷类型并查询，如图 7－19 所示。

```
[root@ controller bin] # cinder type - create lvm
[root@ controller bin] # cinder type - list
```

```
[root@controller bin]# cinder type-create lvm
+--------------------------------------+------+-------------+-----------+
| ID                                   | Name | Description | Is_Public |
+--------------------------------------+------+-------------+-----------+
| 888da805-bd61-478d-836d-91bbac06aaa9 | lvm  | -           | True      |
+--------------------------------------+------+-------------+-----------+
[root@controller bin]# cinder type-list
+--------------------------------------+------+-------------+-----------+
| ID                                   | Name | Description | Is_Public |
+--------------------------------------+------+-------------+-----------+
| 888da805-bd61-478d-836d-91bbac06aaa9 | lvm  | -           | True      |
+--------------------------------------+------+-------------+-----------+
[root@controller bin]#
```

图 7－19　创建 lvm 标识的卷类型

b. 创建一个带 lvm 标识的云硬盘，如图 7－20 所示。

```
[root@ controller bin] # cinder create - - display - name type_test_demo - - volume_type lvm 1
```

```
[root@controller bin]# cinder create --display-name type_test_demo --volume_type lvm 1
+--------------------------------+---------------------------------------+
| Property                       | Value                                 |
+--------------------------------+---------------------------------------+
| attachments                    | []                                    |
| availability_zone              | nova                                  |
| bootable                       | false                                 |
| consistencygroup_id            | None                                  |
| created_at                     | 2019-08-29T12:26:37.000000            |
| description                    | None                                  |
| encrypted                      | False                                 |
| id                             | 808970bb-eb09-4001-9fef-a368ac726192  |
| metadata                       | {}                                    |
| migration_status               | None                                  |
| multiattach                    | False                                 |
| name                           | type_test_demo                        |
| os-vol-host-attr:host          | compute@lvm#LVM                       |
| os-vol-mig-status-attr:migstat | None                                  |
| os-vol-mig-status-attr:name_id | None                                  |
| os-vol-tenant-attr:tenant_id   | 0fe13ac26d114f9fa91c29e02f09fbd1      |
| replication_status             | disabled                              |
| size                           | 1                                     |
| snapshot_id                    | None                                  |
| source_volid                   | None                                  |
| status                         | creating                              |
| updated_at                     | 2019-08-29T12:26:37.000000            |
| user_id                        | 934a974a70d947949e8790d7442a4e37      |
| volume_type                    | lvm                                   |
+--------------------------------+---------------------------------------+
[root@controller bin]#
```

图 7－20　创建一个带 lvm 标识的云硬盘

④创建并查看 Cinder 快照信息，如图 7 - 21 所示。

[root@ controller bin] # cinder snapshot - create - - display - name snapshot_demo type_test_demo

[root@ controller bin] # cinder list

```
[root@controller bin]#  cinder snapshot-create --display-name snapshot_demo  type_test_demo
+-------------+--------------------------------------+
|   Property  |                Value                 |
+-------------+--------------------------------------+
|  created_at |      2019-08-29T12:27:09.401866      |
| description |                 None                 |
|      id     | ce862d06-e8d4-4612-bf42-928026f9e70c |
|   metadata  |                  {}                  |
|     name    |            snapshot_demo             |
|     size    |                  1                   |
|    status   |               creating               |
|  updated_at |                 None                 |
|  volume_id  | 808970bb-eb09-4001-9fef-a368ac726192 |
+-------------+--------------------------------------+
[root@controller bin]# cinder list
+--------------------------------------+-----------+--------------------+------+-------------+----------+-------------+
|                  ID                  |   Status  |        Name        | Size | Volume Type | Bootable | Attached to |
+--------------------------------------+-----------+--------------------+------+-------------+----------+-------------+
| 6273c549-a274-419c-b021-4cb1521fc644 | available |  cinder-volume-demo |  10  |      -      |  false   |             |
| 808970bb-eb09-4001-9fef-a368ac726192 | available |   type_test_demo   |  1   |     lvm     |  false   |             |
| bc605848-a9dd-4af7-ba66-3a2744b40103 | available | cinder-volume-demo2 |  50  |      -      |  false   |             |
+--------------------------------------+-----------+--------------------+------+-------------+----------+-------------+
[root@controller bin]#
```

图 7 - 21　创建并查看 Cinder 快照信息

5. 在 Dashboard 界面上完成块存储任务

对 Cinder 后端逻辑卷进行扩容。

①请先删除原来的云硬盘，然后创建一个 10 GB 的云硬盘并查看，如图 7 - 22 和图 7 - 23 所示。

创建云硬盘

云硬盘名称：test

描述

云硬盘来源：没有源，空白云硬盘。

类型：未设置云硬盘类型

大小(GiB) *：10

可用域：nova

说明：

云硬盘是可被连接到云主机的块设备。

云硬盘类型描述：

如果"无云硬盘类型"被选中，云硬盘被创建的时候将没有云硬盘类型。

云硬盘限度

总大小 (GB) (0 GiB)　1000 GiB 可用

云硬盘数量 (0)　10 可用

取消　创建云硬盘

图 7 - 22　创建一个 10 GB 的云硬盘 1

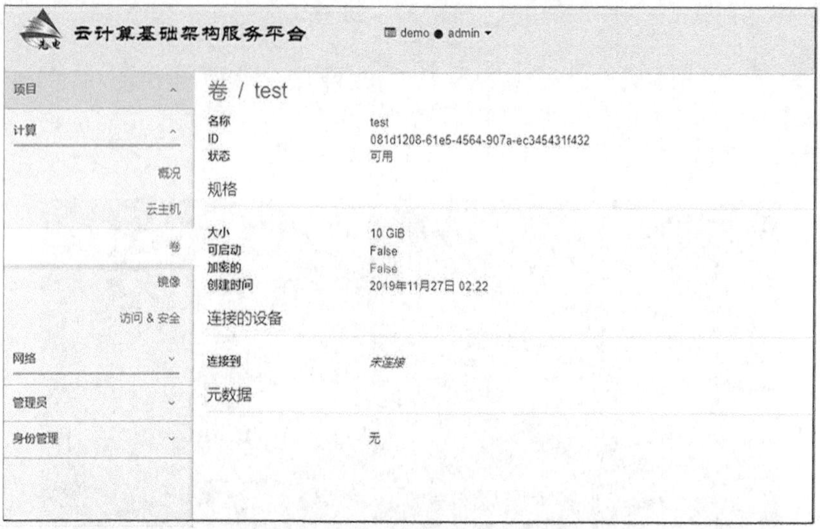

云计算基础架构服务平台
demo ● admin
项目
计算
概况
云主机
卷
镜像
访问 & 安全
网络
管理员
身份管理
卷 / test
名称 test
ID 081d1208-61e5-4564-907a-ec345431f432
状态 可用
规格
大小 10 GiB
可启动 False
加密的 False
创建时间 2019年11月27日 02:22
连接的设备
连接到 未连接
元数据
无

图 7－23　创建一个 10 GB 的云硬盘 2

②扩容至 50 GB，如图 7－24 和图 7－25 所示。

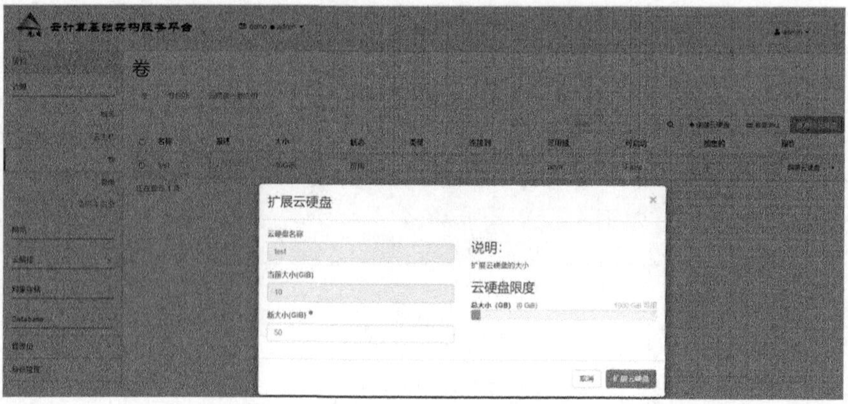

云计算基础架构服务平台
卷
扩展云硬盘
云硬盘名称
test
当前大小(GiB)
10
新大小(GiB) *
50
说明:
扩展云硬盘的大小
云硬盘限度
取消
扩展云硬盘

图 7－24　扩容至 50 GB1

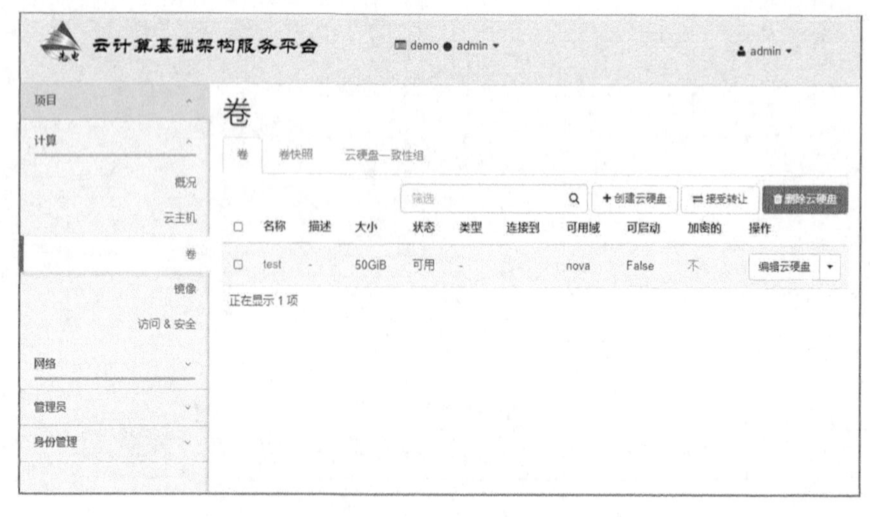

图 7－25　扩容至 50 GB2

③指定 Cinder 卷类型。

a. 创建 type 标识的卷类型并查询，如图 7－26 所示。

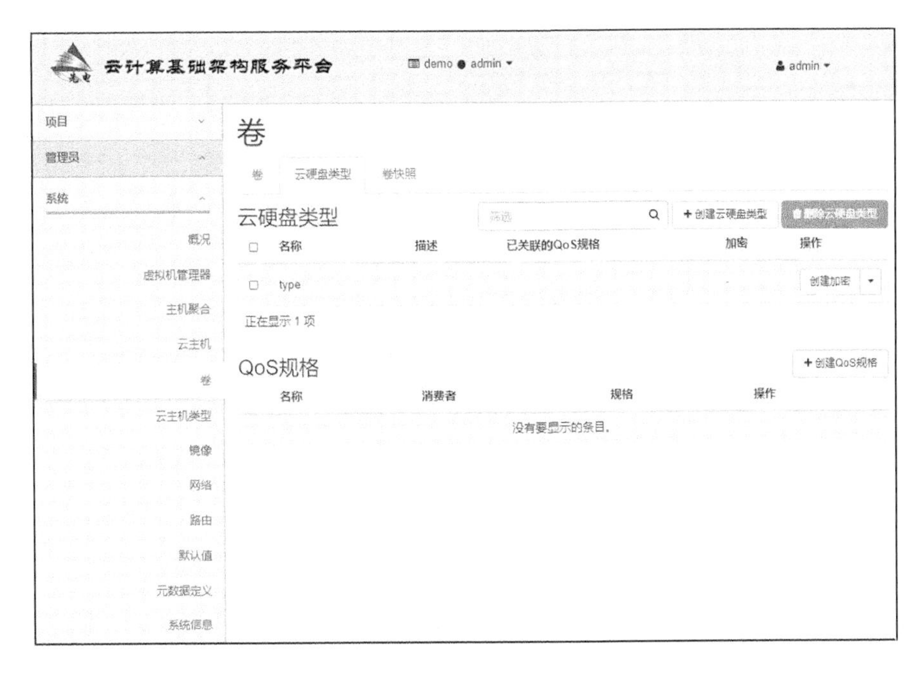

图 7－26　创建 type 标识的卷类型

b. 创建一个带 type 标识的云硬盘，如图 7－27 和图 7－28 所示。

图 7－27　创建一个带 type 标识的云硬盘 1

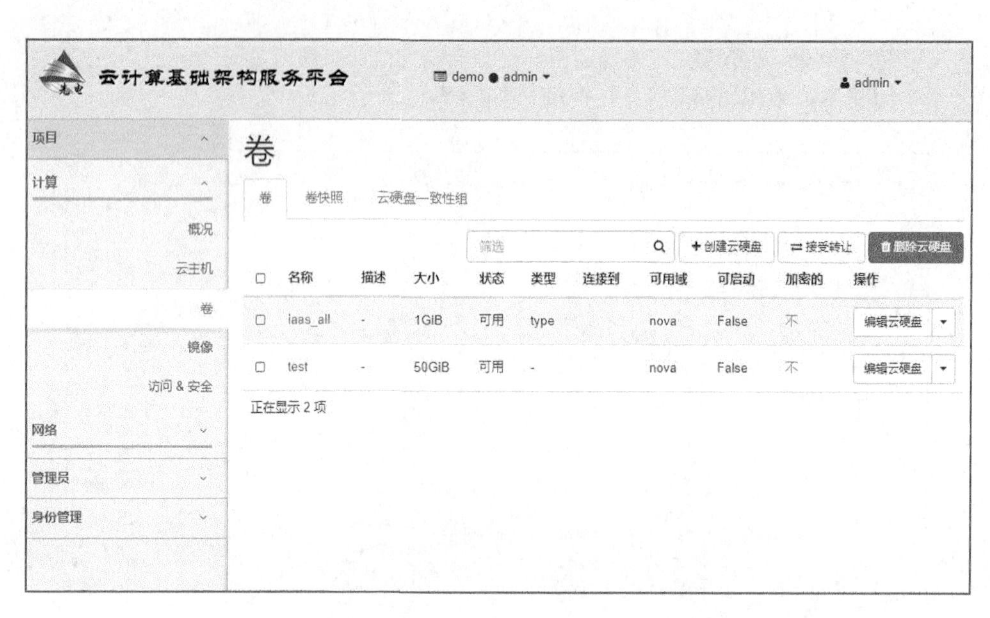

图 7-28　创建一个带 type 标识的云硬盘 2

④为云硬盘 iaas_all 创建快照，名称为 kz，并查看快照信息，如图 7-29 所示。

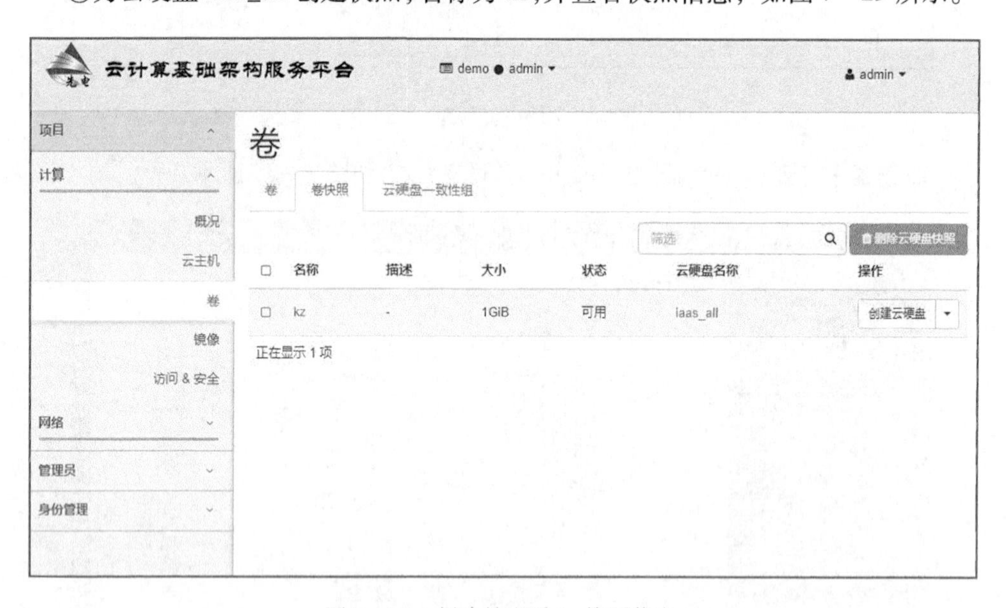

图 7-29　创建并查看 kz 快照信息

任务二　安装与操作对象存储服务

任务要求

现公司内部拥有 3 个用户组，3 个用户组需要内部存储服务器作为内部存储资源，需要每个租户创建以自己名称命名的存储容器、创建后缀为“_Public”的对外服务的存储，创建后缀为“_Private”自己使用的容器。

为保证内部数据的安全性，将 Glance、Cinder 的后端存储修改为 Swift。

核心概念

对象存储服务：OpenStack 存储服务 Swift 的功能是提供高可用分布式对象存储的服务，为 Nova 组件提供虚拟机镜像存储服务。Swift 构筑在比较便宜的标准硬件存储基础设施之上，无须采用 RAID（磁盘冗余阵列），通过在软件层面引入一致性散列技术和数据冗余性，牺牲一定程度的数据一致性来达到高可用性和可伸缩性，支持多租户模式、容器和对象读/写操作，适合解决互联网的应用场景下非结构化数据存储问题。

视频

对象存储的相关知识

知识准备

1. Swift 的来源

Swift 最初是由 Rackspace 公司开发的高可用分布式对象存储服务，并于 2010 年贡献给 OpenStack 开源社区作为其最初的核心子项目之一，为其 Nova 子项目提供虚拟机镜像存储服务。Swift 构筑在比较便宜的标准硬件存储基础设施之上，无须采用 RAID（磁盘冗余阵列），通过在软件层面引入一致性散列技术和数据冗余性，牺牲一定程度的数据一致性来达到高可用性和可伸缩性，支持多租户模式、容器和对象读/写操作，适合解决互联网的应用场景下非结构化数据存储问题。此项目是基于 Python 开发的，采用 Apache 2. 0 许可协议，可用来开发商用系统。Swift 为 OpenStack 提供一种分布式、持续虚拟对象存储，它类似于 Amazon Web Service 的 S3 简单存储服务。Swift 具有跨节点百级对象的存储能力。Swift 内建冗余和失效备援管理，也能够处理归档和媒体流，特别是对大数据（千兆字节）和大容量（多对象数量）的测度非常高效。

2. Swift 的主要特征

（1）极高的数据持久性（Durability）。

（2）完全对称的系统架构："对称" 意味着 Swift 中各节点可以完全对等，能极大地降低系统维护成本。

（3）无限的可扩展性：包括两方面的意义，一是数据存储容量无限可扩展；二是 Swift 性能（如 QPS、吞吐量等）可线性提升。

（4）无单点故障：Swift 的元数据存储是完全均匀随机分布的，并且与对象文件存储一样，元数据也会存储多份。整个 Swift 集群中，也没有一个角色是单点的，并且在架构和设计上保证无单点业务是有效的。

（5）简单、可依赖。

3. Swift 的组件

Swift 组件包括：Swift 代理、Swift 对象、Swift 容器、Swift 账户、Swift RING。

（1）Swift 代理服务器。用户都是通过 Swift - API 与代理服务器进行交互，代理服务器正是接收外界请求的门卫，它检测合法的实体位置并路由它们的请求。此外，代理服务器也同时处理实体失效而转移时，故障切换的实体重复路由请求。

（2）Swift 对象服务器。对象服务器是一种二进制存储，它负责处理本地存储中的对象数据的存储、检索和删除。对象都是文件系统中存放的典型的二进制文件，具有扩展文件属性的元数据（xattr）。

注：xattr 格式被 Linux 中的 ext3/4、XFS、Btrfs、JFS 和 ReiserFS 所支持，但是并没有有效测试证明在 XFS、JFS、ReiserFS、Reiser4 和 ZFS 下也同样能运行良好。不过，XFS 被认为是当前最好的选择。

（3）Swift 容器服务器。容器服务器列出一个容器中的所有对象，默认对象列表将存储为 SQLite 文件（也可以修改为 MySQL，安装中就是以 MySQL 为例）。容器服务器也会统计容器中包含的对象数量及容器的存储空间耗费。

（4）Swift 账户服务器。账户服务器与容器服务器类似，将列出容器中的对象。

（5）Swift Ring（索引环）。Ring 容器记录着 Swift 中物理存储对象的位置信息，它是真实物理存储位置的实体名的虚拟映射，类似于查找及定位不同集群的实体真实物理位置的索引服务。这里所谓的实体指账户、容器、对象，它们都拥有属于自己的不同的 Rings。

1. 准备实施任务的环境（见表 7－3）

对象存储的基本操作

表 7－3 实施任务的准备环境

实施任务所需软件资源	虚拟机镜像资源信息
虚拟机软件（vm12）	controller
securcrt 远程连接软件	compute

2. 分解实施任务的过程（见表 7－4）

表 7－4 实施任务的简明过程

序号	步 骤	详细操作及说明
1	对象服务的基本操作	①安装 Swift 组件； ②Swift 组件的基本操作
2	创建符合公司要求的对象存储任务	①在 Dashboard 界面创建容器； ②通过 CLI 命令行创建容器

3. 对象服务 Swift 基本操作

（1）安装 Swift 组件。

①导入环境变量。

```
[root@ controller ~] #source /etc/keystone/admin-openrc.sh
```

②在 controller 节点上先挂载 iso，如果/opt 有镜像文件就无须挂载再修改环境变量，如图 7－30 所示。

```
[root@ controller ~] #mount -o loop /opt/CentOS-7-x86_64-DVD-1511.iso /opt/centos7
[root@ controller ~] #mount -o loop /opt/XianDian-IaaS-v2.2.iso /opt/iaas
[root@ controller ~] #cd /usr/local/bin
[root@ controller ~] #vi /etc/xiandian/openrc.sh
```

按【I】键进入编译模式，将文件修改为以下形式：

```
SWIFT_PASS =000000
OBJECT_DISK =sdd
STORAGE_LOCAL_NET_IP =192.168.100.20
```

按【Esc】键并输入：wq 命令保存退出。

```
##--------------------Swift Config---------------------##
##Password for Keystore swift user. exmaple:000000
SWIFT_PASS=000000

##The NODE Object Disk for Swift. example:md126p4.
OBJECT_DISK=sdd

##The NODE IP for Swift Storage Network. example:x.x.x.x.
STORAGE_LOCAL_NET_IP=192.168.100.20

##--------------------Heat Config----------------------##
```

图 7－30　修改 controller 节点的虚拟机上的环境变量

③在控制节点安装 Swift 组件。

```
[root@ controller ~] # cd /usr/local/bin
[root@ controller bin] # iaas - install - swift - controller.sh
```

④在 compute 节点的虚拟机上增加一个 50 GB 的新硬盘，重启并且修改环境变量，与控制节点的配置一样，如图 7－31 和图 7－32 所示。

```
[root@ compute ~] # vi /etc/xiandian/openrc.sh
```

图 7－31　在 compute 节点的虚拟机上增加一个 50 GB 的新硬盘

按【I】键进入编译模式，将文件修改为以下形式：

```
SWIFT_PASS=000000
OBJECT_DISK=sdd
STORAGE_LOCAL_NET_IP=192.168.100.20
```

```
##--------------------Swift Config--------------------##
##Password for Keystore swift user. exmaple:000000
SWIFT_PASS=000000

##The NODE Object Disk for Swift. example:md126p4.
OBJECT_DISK=sdd

##The NODE IP for Swift Storage Network. example:x.x.x.x.
STORAGE_LOCAL_NET_IP=192.168.100.20
```

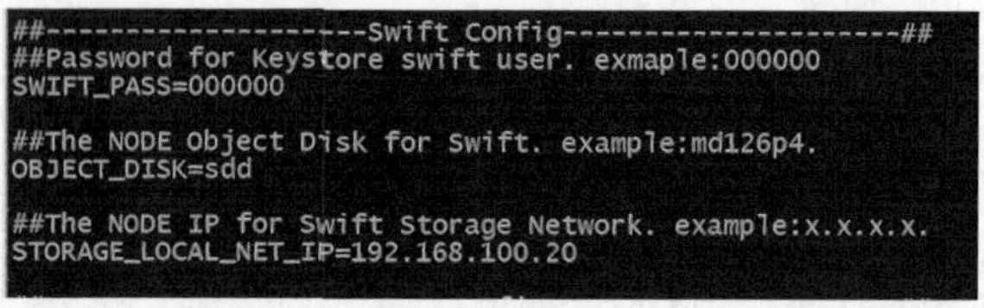

图 7-32　修改 compute 节点的虚拟机上的环境变量

⑤安装 iaas-install-swift-compute.sh 脚本。

```
[root@compute bin]# cd /usr/local/bin
[root@compute bin]# iaas-install-swift-compute.sh
```

⑥验证 Swift 组件是否安装成功，如图 7-33 所示。

```
[root@controller bin]# swift stat
        Account: AUTH_26b08dcad5a74c09b1ad0fda0a611bce
     Containers: 0
        Objects: 0
          Bytes: 0
X-Put-Timestamp: 1567088293.85264
    X-Timestamp: 1567088293.85264
     X-Trans-Id: tx6e4f7e730c0848dba9243-005d67dea3
   Content-Type: text/plain; charset=utf-8
[root@controller bin]#
```

图 7-33　验证 Swift 组件是否安装成功

（2）Swift 基本操作。在 controller 节点上操作。

①通过命令行实现对 Swift 上数据的操作，首先需要创建一个名为 xiandian 的容器，如图 7-34 所示。

```
[root@controller bin]# swift post xiandian
```

```
[root@controller bin]# swift post xiandian
[root@controller bin]#
```

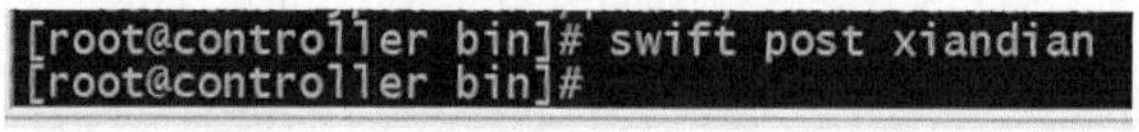

图 7-34　创建一个名为 xiandian 的容器

②查看创建窗口是否成功，如图 7-35 所示。

```
[root@controller bin]# swift stat
```

```
[root@controller bin]# swift stat
                        Account: AUTH_26b08dcad5a74c09b1ad0fda0a611bce
                     Containers: 1
                        Objects: 0
                          Bytes: 0
Containers in policy "policy-0": 1
   Objects in policy "policy-0": 0
     Bytes in policy "policy-0": 0
   X-Account-Project-Domain-Id: 5b68813be9364902bf40477cb7f67d1d
                    X-Timestamp: 1567088404.06539
                     X-Trans-Id: tx5cc959998a764f648def8-005d67df52
                   Content-Type: text/plain; charset=utf-8
                  Accept-Ranges: bytes
[root@controller bin]#
```

图 7-35　查看创建窗口是否成功

③创建 test 目录，并将本地的 test 目录内容递归上传到 xiandian 容器内，上传时首先需要上传一个空白的 test 目录，如图 7－36 和图 7－37 所示。

```
[root@ controller bin] #mkdir test
[root@ controller ~] # swift upload xiandian test
```

```
[root@controller bin]#  mkdir test
[root@controller bin]#
```

图 7－36　创建 test 目录

```
[root@controller bin]# swift upload xiandian test
test
```

图 7－37　将本地的 test 目录内容递归上传到 xiandian 容器

④有了容器之后，可以查看 xiandian 容器中的内容，如图 7－38 所示。

```
[root@ controller ~] #swift list xiandian
```

```
[root@controller bin]# swift list xiandian
test
```

图 7－38　查看 xiandian 容器中的内容

⑤将 iaas. txt 文件上传到 xiandian 容器的 test 目录内，如图 7－39 所示。

```
[root@ controller bin] # cd test
[root@ controller test] # touch iaas.txt
[root@ controller test] # swift upload xiandian/test
iaas.txt
```

```
[root@controller bin]# cd test
[root@controller test]# touch iaas.txt
[root@controller test]#  swift upload xiandian/test iaas.txt
test/iaas.txt
[root@controller test]#
```

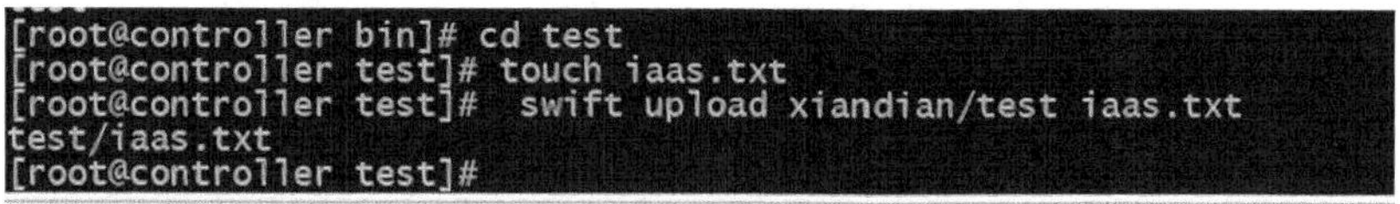

图 7－39　将 iaas. txt 文件上传到 xiandian 容器的 test 目录

⑥数据在 swift 集群内保存，随时供用户下载使用，现在主目录下载 iaas. txt 文件，如图 7－40 所示。

```
[root@ controller ~] # cd /tmp
[root@ controller tmp] #swift download xiandian test/iaas.txt
[root@ controller tmp] # ls
```

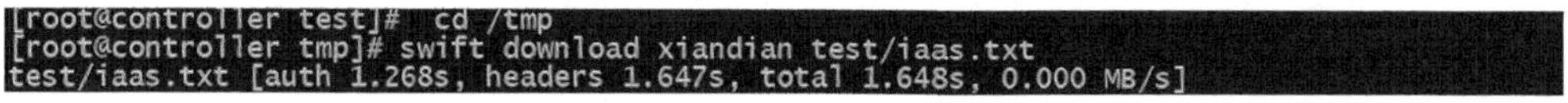

```
[root@controller test]#  cd /tmp
[root@controller tmp]# swift download xiandian test/iaas.txt
test/iaas.txt [auth 1.268s, headers 1.647s, total 1.648s, 0.000 MB/s]
```

图 7－40　数据在 swift 集群内保存，随时供用户下载使用

⑦磁盘容量有限，需要删除一些相对价值低的数据空出更多的空间，如图 7－41 所示。

```
[root@ controllertmp] # swift delete xiandian test/iaas.txt
```

```
[root@controller tmp]#  swift delete xiandian test/iaas.txt
test/iaas.txt
```

图 7-41　磁盘容量有限，删除一些数据

⑧通过 swift stat 命令来首先查看 Account 账户下 swift 状态，如图 7-42 所示。

```
[root@ controller tmp] #swift stat
```

```
[root@controller tmp]# swift stat
                        Account: AUTH_26b08dcad5a74c09b1ad0fda0a611bce
                     Containers: 1
                        Objects: 1
                          Bytes: 0
Containers in policy "policy-0": 1
   Objects in policy "policy-0": 1
     Bytes in policy "policy-0": 0
    X-Account-Project-Domain-Id: 5b68813be9364902bf40477cb7f67d1d
                    X-Timestamp: 1567088404.06539
                     X-Trans-Id: tx7f62fa0aa8e74ac3844a1-005d67e017
                   Content-Type: text/plain; charset=utf-8
                  Accept-Ranges: bytes
[root@controller tmp]#
```

图 7-42　查看 Account 账户下 swift 状态

⑨可以查看具体容器的运行状态，以查看 xiandian 容器为例，如图 7-43 所示。

```
[root@ controller tmp] #swift stat xiandian
```

```
[root@controller tmp]# swift stat xiandian
          Account: AUTH_26b08dcad5a74c09b1ad0fda0a611bce
        Container: xiandian
          Objects: 1
            Bytes: 0
         Read ACL:
        Write ACL:
          Sync To:
         Sync Key:
    Accept-Ranges: bytes
 X-Storage-Policy: Policy-0
      X-Timestamp: 1567088404.09717
       X-Trans-Id: txf6592f5e76e143e69c4e1-005d67e029
     Content-Type: text/plain; charset=utf-8
```

图 7-43　查看 xiandian 容器的运行状态

⑩可以查看 xiandian 容器内具体某个对象 test 的状态，如图 7-44 所示。

```
[root@ controller tmp] #swift stat xiandian test
```

```
[root@controller tmp]#  swift stat xiandian test
       Account: AUTH_26b08dcad5a74c09b1ad0fda0a611bce
     Container: xiandian
        Object: test
  Content Type: text/directory
Content Length: 0
 Last Modified: Thu, 29 Aug 2019 14:22:09 GMT
          ETag: d41d8cd98f00b204e9800998ecf8427e
    Meta Mtime: 1567088508.303960
 Accept-Ranges: bytes
   X-Timestamp: 1567088528.62614
    X-Trans-Id: tx63480f9911f7404288220-005d67e03b
```

图 7-44　查看 xiandian 容器内的对象 test 的状态

4. 创建符合公司要求的对象存储任务

（1）在 Dashboard 界面创建容器。

为项目研发部创建公共存储容器，名为 RD_Dept_Public，如图 7-45和图 7-46 所示。

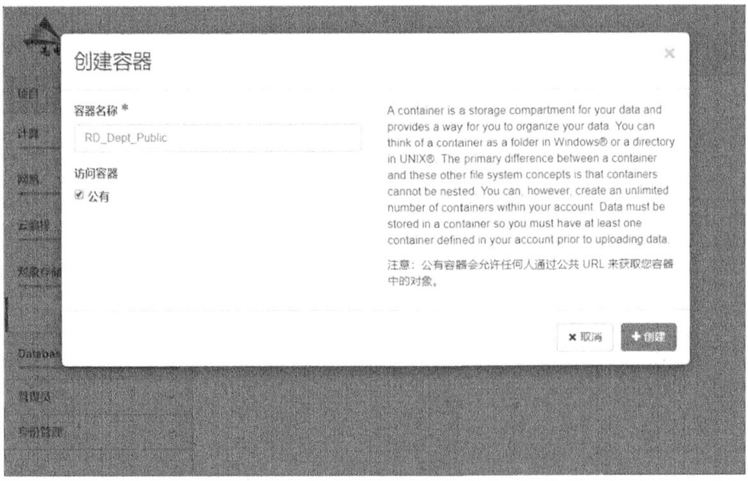

图 7-45　为项目研发部创建公共存储容器 1

图 7-46　为项目研发部创建公共存储容器 2

（2）通过 CLI 命令行创建容器。

①为业务部创建私有存储容器，名为 BS_Dept_Private，如图 7-47 所示。

```
[root@controller ~]# swift post BS_Dept_Private
```

图 7-47　为业务部创建私有存储容器

②为 IT 工程部创建私有存储容器，名为 IT_Dept_Private，如图 7-48 所示。

图 7-48　为 IT 工程部创建私有存储容器

项目小结

Cinder 作为 OpenStack 的块存储服务，Swift 是 OpenStack 的对象存储服务，为虚拟机提供虚拟磁盘。本项目首先学习 Cinder 和 Swift 的架构，然后讨论了 Cinder 和 Swift 的各个服务组件，最后通过两个任务实施使读者完成项目目标。

表7－5列举本项目Cinder和Swift的常用命令及其说明。

表7－5　常用命令及其说明

常用命令	命令说明
cinder list	显示存储卷列表
cinder type－list	显示存储卷类型列表
cinder create －－display－name VOLNAME SIZE	创建存储卷
cinder extend VOLUME_ ID SIZE_ IN_ GB	重置卷设备大小
cinder delete VOLNAME－OR－ID	删除存储卷
cinder rename VOLNAME－OR－ID NEW－VOLNAME	重命名存储卷
cinder show VOLNAME－OR－ID	显示存储卷信息
cinder snapshot－create －－display－name SNAPSHOT－VOLNAME VOLNAME－OR－ID	创建存储卷快照
cinder snapshot－list	显示存储卷快照列表
swift list	列举容器的内容
swift post	创建容器
swift stat	查看容器的运行状态
swift delete	删除容器的某些内容
swift upload	上传文件到容器

结合本项目，思考Cinder如何创建卷（volume），它的过程包括哪些方面？

项目八 高级控制服务

项目综述

小张在学习了云平台所需的基本组件后，继续学习辅助平台管理和生产要求的一些高级的平台组件，提高平台快速部署实例的应用能力及提升平台的监控管理能力。小张为此要完成的任务如下：

视频

项目八综述

- 掌握Heat模板的编写方法。
- 利用Heat模板实现实例的批量自动部署及应用。
- 利用监控服务实时查看云平台的运行情况。

项目目标

【知识目标】

- OpenStack的Heat服务组件的相关概念。
- OpenStack的Heat编配服务流程和工作机制。

【技能目标】

- 掌握Heat模板的编写方法。
- 利用Heat模板实现实例的批量自动部署及应用。
- 利用监控服务实时查看平台的运行情况。

【职业能力】

根据企业的应用需求，利用Heat模板实现实例的批量自动部署和利用监控服务实时查看云平台的运行情况。

任务一　安装与配置编配服务

任务要求

Heat模板文件可以实现实例的批量自动部署及应用，保证用户正常使用云平台实例资源。编写Heat模板文件时可根据需要选择相应的镜像和网络。

小张要为项目研发部门、业务部门和IT工程部门编写Heat模板文件，该模板文件可以启动3台实例，其中一台为Windows 7 64位系统、4 GB内存和60 GB的磁盘空间，另外两台为CentOS 7.2操作系统，1 GB内存和50 GB的磁盘空间。

核心概念

编配服务：OpenStack编配服务Heat是一个基于模板来编排复合云应用的服务。Heat向开发人员和系统管理员提供了一种简便地创建和管理一批相关的OpenStack资源的方法，并通过有序且可预测的方式对其进行资源配置和更新。

知识准备

1. 基本概念

编排，顾名思义，就是按照一定的目的依次排列。在IT世界，一个完整的编排一般包括设置服务器的机器、安装CPU、内存、硬盘、通电、插入网络接口、安装操作系统、配置操作系统、安装中间件、配置中间件、安装应用程序以及配置应用发布程序。对于复杂的如图8－1所示的Heat架构需要部署在多台服务器上的应用，需要重复这个过程，而且需要协调各个应用模块的配置。OpenStack以命令行和Horizon的方式提供给用户进行资源管理，但是这两种方法的工作效率并不高。即使把命令行保存为脚本，在输入/输出，依赖关系之间仍需要编写额外的脚本来进行维护，而且不易于扩展。如果用户直接通过REST API编写程序，同样会引发额外的复杂性。因此，这两种方式都不利于用户通过OpenStack进行批量资源管理和编排各种资源。

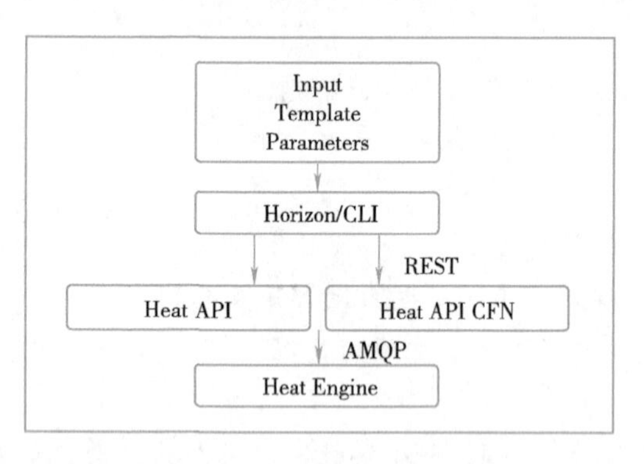

图8－1　Heat架构

Heat服务包含以下重要的组件：

① Heat API组件实现OpenStack天然支持的REST API。该组件通过把API请求经由AMQP传送给Heat Engine来处理API请求。

② Heat API CFN组件提供兼容AWS CloudFormation的API，同时也会把API请求通过AMQP转发给Heat Engine。

③ Heat Engine组件提供Heat最主要的协作功能。

2. Heat的重要性

1）更快更有效地管理OpenStack的资源

云平台系统在相对比较稳定的情况下，管理成本逐渐变成首要需要解决的问题。

云上自动化能力是一个云平台的刚需，可以有效降低维护难度。

OpenStack 原生提供命令行和 Horizon 来供用户管理资源。然而命令行和在浏览器中的单击操作都费时费力，不利于用户使用 OpenStack 进行大批量的管理以支撑 IT 应用。Heat 在这种情况下应运而生。

Heat 采用了模板方式来设计或者定义编排。为方便用户使用，Heat 还提供了大量的模板例子，使用户能够方便地得到想要的编排。

2）更小的研发成本

引入 Heat，对于不了解 OpenStack 的研发者来说，可以更快地接入现有的业务系统。开发者更关心的是授权认证和对虚拟资源的增加、删除、修改，而对于底层的状态并不用太多了解。

3. 基本术语

1）Stack

Stack 概念来源于 AWS，是 OpenStack 中用来管理一组资源的基本单位。一个 stack 往往对应一个应用程序。Stack 管理的是 resource，而 resource 是抽象的概念，它可以是虚拟机、网络等。Stack 就是在单个模板中定义的实例化资源的集合，是 Heat 管理应用程序的逻辑单元。

2）template

Heat 的 template 描述了所用的所有组件资源以及组件资源之间的关系。heat 模版是 heat 的核心。

3）resources

资源是底层服务的抽象，CPU、memory、disk、网络等都可以看作是资源。一个 stack 可以拥有很多资源。资源和资源之间会存在依赖关系。Heat 在创建栈时会自动解析依赖关系，按顺序创建资源。在 Heat 的 template 中，resources 用于模板中资源的声明，在 HOT 模板中，应该至少有一个资源的定义，否则在实例化模板时将不会做任何事情。

4）parameters

Heat 模板中的顶级 key，定义在创建或更新 stack 时可以传递哪些数据来定制模板。

5）parameter_groups

用于指定如何对输入参数进行分组，以及提供参数的顺序。

6）@ outputs

Heat 模板中的顶级 key，定义实例化后 stack 将返回的数据。

顶级 key 包括七个：heat_template_version、description、parameter_groups、parameters、resources、outputs、conditions。除了 heat_template_version 和 resources，其他都是可选部分。

4. Heat 模板

Heat 支持两种格式的模板：一种是基于 JSON 格式的 CFN 模板，另外一种是基于 YAML 格式的 HOT 模板。CFN 模板主要是为了保持对 AWS 的兼容性。HOT 模板是 Heat 自有的，资源类型更加丰富，更能体现 Heat 特点的模板。

一个典型的 HOT 模板由下列元素构成：

① 模板版本必填字段，指定所对应的模板版本，Heat 会根据版本进行检验。

② 参数列表选填，指输入参数列表。

③ 资源列表必填，指生成的 Stack 所包含的各种资源。可以定义资源间的依赖关系，如生成 Port，然后再用 Port 来生成 VM。

④ 输出列表选填，指生成的 Stack 暴露的信息，可以用来给用户使用，也可以用来作为输入提供给其他的 Stack。

对于不同的资源，Heat 都提供了对应的资源类型。可以通过 Heat 的命令进行查询。执行命令如下：

```
[root@ controller ~] # source /etc/keystone/admin-openrc.sh
[root@ controller ~] # heat resource-type-list
```

执行返回如图 8-2 所示。

```
[root@xiandian ~]# source /etc/keystone/admin-openrc.sh
[root@xiandian ~]# heat resource-type-list
+------------------------------------------------+
| resource_type                                  |
+------------------------------------------------+
| AWS::AutoScaling::AutoScalingGroup             |
| AWS::AutoScaling::LaunchConfiguration          |
| AWS::AutoScaling::ScalingPolicy                |
| AWS::CloudFormation::Stack                     |
| AWS::CloudFormation::WaitCondition             |
| AWS::CloudFormation::WaitConditionHandle       |
| AWS::CloudWatch::Alarm                         |
| AWS::EC2::EIP                                  |
| AWS::EC2::EIPAssociation                       |
| AWS::EC2::Instance                             |
| AWS::EC2::InternetGateway                      |
| AWS::EC2::NetworkInterface                     |
| AWS::EC2::RouteTable                           |
| AWS::EC2::SecurityGroup                        |
| AWS::EC2::Subnet                               |
| AWS::EC2::SubnetRouteTableAssociation          |
| AWS::EC2::VPC                                  |
| AWS::EC2::VPCGatewayAttachment                 |
| AWS::EC2::Volume                               |
| AWS::EC2::VolumeAttachment                     |
| AWS::ElasticLoadBalancing::LoadBalancer        |
| AWS::IAM::AccessKey                            |
| AWS::IAM::User                                 |
| AWS::RDS::DBInstance                           |
| AWS::S3::Bucket                                |
| OS::Ceilometer::Alarm                          |
| OS::Ceilometer::CombinationAlarm               |
| OS::Ceilometer::GnocchiAggregationByMetricsAlarm |
| OS::Ceilometer::GnocchiAggregationByResourcesAlarm |
| OS::Ceilometer::GnocchiResourcesAlarm          |
| OS::Cinder::EncryptedVolumeType                |
| OS::Cinder::Volume                             |
| OS::Cinder::VolumeAttachment                   |
| OS::Cinder::VolumeType                         |
| OS::Glance::Image                              |
| OS::Heat::AccessPolicy                         |
| OS::Heat::AutoScalingGroup                     |
| OS::Heat::CloudConfig                          |
| OS::Heat::HARestarter                          |
| OS::Heat::InstanceGroup                        |
| OS::Heat::MultipartMime                        |
| OS::Heat::None                                 |
| OS::Heat::RandomString                         |
| OS::Heat::ResourceChain                        |
| OS::Heat::ResourceGroup                        |
| OS::Heat::ScalingPolicy                        |
| OS::Heat::SoftwareComponent                    |
```

图 8-2 Heat 资源类型列表

5. Heat 功能描述

Heat 编排服务，通过使用描述性的模板格式，来编排复合云应用程序。

（1）Heat 提供基于模板的业务流程，调用相应的 OpenStack API，生成运行的云应用程序。

（2）Heat 模板在文本文件中，描述了云应用程序的基础结构，这些文本文件可读可写，并且可以通过版本控制工具进行管理。

（3）模板指定了资源之间的关系（如指定存储卷连接到指定服务器），这使 Heat 能够调用 OpenStack API，以正确的顺序创建所有基础设施，启动应用程序。

（4）Heat 集成了 OpenStack 的其他组件，能够自动化调配大多数云资源（如实例、浮动 IP、卷、安全组、用户等），以及一些更高级的功能，如高可用性、实例自动伸缩和嵌套 stack。

（5）Heat 模板与软件配置管理工具（如 Puppet 和 Ansible）的集成。

（6）操作人员可以通过安装插件定制 Heat 功能。

编配服务的基本操作

1. 准备实施任务的环境（见表 8－1）

表 8－1 实施任务的准备环境

实施任务所需软件资源	虚拟机镜像资源信息
虚拟机软件（vm12）	controller
securcrt 远程连接软件	compute

2. 分解实施任务的过程（见表 8－2）

表 8－2 实施任务的简明过程

序号	步　骤	详细操作及说明
1	编配服务的基本操作	①安装编配服务； ②使用栈模板 test－stack. yml 创建一个名为 Orchestration 的栈； ③查看栈列表； ④查看栈的详细信息； ⑤查看栈资源； ⑥查看栈输出列表； ⑦查看栈输出值； ⑧查看栈事件列表； ⑨查看栈资源事件详细信息； ⑩删除栈
2	创建符合企业要求的编配任务	①编写 RD_Dept. yml 文件； ②通过 Dashboard 为研发部启动栈资源，创建实例

3. 安装编配服务（见图 8－3）

```
[root@controller bin]# iaas-install-heat.sh
Loaded plugins: fastestmirror
```

图 8－3 安装编配服务

4. Heat 运维基础操作

(1) 安装编配服务。

```
[root@ controller ~] #source /etc/keystone/admin -openrc. sh
[root@ controller ~] #vi /etc/xiandian/openrc. sh
```

按【I】键进入编译模式，将文件修改为以下形式：

```
HEAT_ DBPASS =000000
HEAT_ PASS =000000
```

按【Esc】键并输入：wq 命令保存退出。

安装 Heat 组件：

```
[root@ controller ~] #iaas -install -heat. sh
```

(2) 使用栈模板 test - stack. yml 创建一个名为 Orchestration 的栈。

检测模板文件用到的镜像文件名称和云主机类型及可用网络，并设置 NET_ ID 环境变量表示网络 ID 执行命令如下：

```
[root@ controller ~] # glance image -list
[root@ controller ~] # openstack network list
[root@ controller ~] #openstack flavor list
[root@ controller ~] # export NET_ ID = $ (openstack network
list | awk '/int -net /{ print $2 }')
```

执行返回结果如图 8 - 4 ~ 图 8 - 6 所示。

```
[root@controller ~]# glance image-list
+--------------------------------------+--------------+
| ID                                   | Name         |
+--------------------------------------+--------------+
| 924d8136-c3b3-4d4f-8240-ab837c9afba5 | Centos6.5    |
| e9a5f197-ccef-4f9b-8015-f6b1881e7c71 | Centos7.2    |
| 7f91523a-5f3a-48ac-9583-371d4069dc5b | Centos7.2-2  |
+--------------------------------------+--------------+
```

图 8 - 4　检测镜像名称

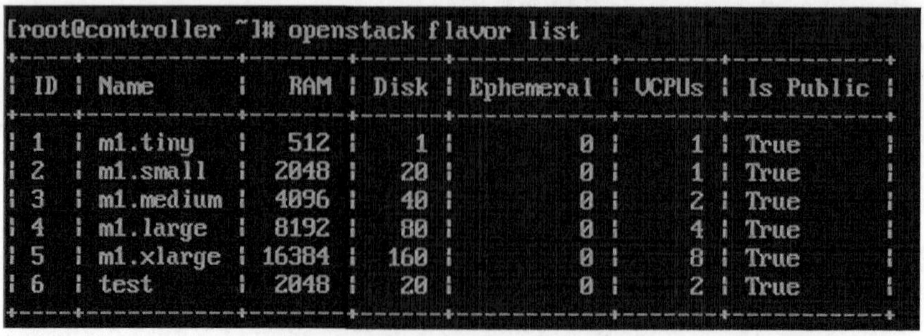

```
[root@controller ~]# openstack flavor list
+----+-----------+-------+------+-----------+-------+-----------+
| ID | Name      |   RAM | Disk | Ephemeral | VCPUs | Is Public |
+----+-----------+-------+------+-----------+-------+-----------+
| 1  | m1.tiny   |   512 |    1 |         0 |     1 | True      |
| 2  | m1.small  |  2048 |   20 |         0 |     1 | True      |
| 3  | m1.medium |  4096 |   40 |         0 |     2 | True      |
| 4  | m1.large  |  8192 |   80 |         0 |     4 | True      |
| 5  | m1.xlarge | 16384 |  160 |         0 |     8 | True      |
| 6  | test      |  2048 |   20 |         0 |     2 | True      |
+----+-----------+-------+------+-----------+-------+-----------+
```

图 8 - 5　检测可用云主机类型

```
[root@controller ~]# openstack network list
+--------------------------------------+---------+--------------------------------------+
| ID                                   | Name    | Subnets                              |
+--------------------------------------+---------+--------------------------------------+
| 7e0b0ebc-dfaf-4c34-b959-f8749fbdefaa | ext-net | f46951c2-a7cd-44f0-9744-beda1ee948b8 |
| 50ecd41e-5ef1-499b-ba7f-acd3fb50f788 | int-net | b1d21b66-7b58-44e2-9a59-9e1076410e8c |
+--------------------------------------+---------+--------------------------------------+
```

图 8-6 检测可用的网络

编辑 test-stack. yml 文件内容，执行命令如下：

```
[root@ controller ~] #vi test-stack. yml
```

按【I】键进入编译模式，输入以下内容：

```
heat_ template_ version: 2015-10-15
description: Launch a basic instance with centos7. 2 image
using the
        \ \m1. small \ \ flavor and one network.
parameters:
  NetID:
    type: string
    description: Network ID to use for the instance.
resources:
  Server1:
    type: OS:: Nova:: Server
    properties:
      image: Centos7. 2
      flavor: m1. small
      networks:
      - network: { get_ param: NetID }
outputs:
  instance_ name:
    description: Name of the instance.
    value: { get_ attr: [ Server1, name ] }
  instance_ ip:
    description: IP address of the instance.
    value: { get_ attr: [ Server1, first_ address ] }
```

按【Esc】键并输入：wq 命令保存退出。

将 test-stack. yml 赋予执行权限，执行命令如下：

```
[root@ controller ~] #chmod +x test-stack. yml
```

创建一个名为 Orchestration 的栈，执行命令如下：

```
[root@ controller ~] #openstack stack create -t test-stack.
yml --parameter "NetID=$NET_ID" orchestration
```

执行返回如图 8 -7 所示。

```
[root@controller ~]# openstack stack create -t test-stack.yml --parameter "NetID=$NET_ID" orchestration
+---------------------+--------------------------------------------------------------------------------+
| Field               | Value                                                                          |
+---------------------+--------------------------------------------------------------------------------+
| id                  | 470ea081-f598-4f47-9ed5-b6dbbe35c92d                                           |
| stack_name          | orchestration                                                                  |
| description         | Launch a basic instance with centos7.2 image using the ``m1.small`` flavor and one network. |
| creation_time       | 2019-11-27T02:46:36                                                            |
| updated_time        | None                                                                           |
| stack_status        | CREATE_IN_PROGRESS                                                             |
| stack_status_reason | Stack CREATE started                                                           |
+---------------------+--------------------------------------------------------------------------------+
```

图 8 -7　创建 Orchestration 栈

(3) 查看栈列表，执行命令如下：

```
[root@ controller ~] # heat stack - list
```

执行返回结果如图 8 -8 所示。

```
[root@controller ~]# heat stack-list
+--------------------------------------+---------------+--------------------+---------------------+--------------+
| id                                   | stack_name    | stack_status       | creation_time       | updated_time |
+--------------------------------------+---------------+--------------------+---------------------+--------------+
| 470ea081-f598-4f47-9ed5-b6dbbe35c92d | orchestration | CREATE_IN_PROGRESS | 2019-11-27T02:46:36 | None         |
+--------------------------------------+---------------+--------------------+---------------------+--------------+
```

图 8 -8　查看栈列表

(4) 查看栈的详细信息，执行命令如下：

```
[root@ controller ~] # heat stack - show orchestration
```

执行返回结果如图 8 -9 所示。

```
[root@controller ~]# heat stack-show orchestration
+-----------------------+------------------------------------------------------------
| Property              | Value
+-----------------------+------------------------------------------------------------
| capabilities          | []
| creation_time         | 2019-11-27T02:46:36
| description           | Launch a basic instance with centos7.2 image using the
|                       | ``m1.small`` flavor and one network.
| disable_rollback      | True
| id                    | 470ea081-f598-4f47-9ed5-b6dbbe35c92d
| links                 | http://controller:8004/v1/d2b88800df664a8cb196ec75034c9e45/stacks/orchestration/470ea081-f598-
| notification_topics   | []
| outputs               | [
|                       |   {
|                       |     "output_value": null,
|                       |     "output_key": "instance_name",
|                       |     "description": "Name of the instance."
|                       |   },
|                       |   {
|                       |     "output_value": null,
|                       |     "output_key": "instance_ip",
|                       |     "description": "IP address of the instance."
|                       |   }
|                       | ]
| parameters            | {
|                       |   "OS::project_id": "d2b88800df664a8cb196ec75034c9e45",
|                       |   "OS::stack_id": "470ea081-f598-4f47-9ed5-b6dbbe35c92d",
|                       |   "OS::stack_name": "orchestration",
|                       |   "NetID": "864c05da-fbb1-4f5d-af0c-621200a9d65e"
|                       | }
| parent                | None
| stack_name            | orchestration
| stack_owner           | None
| stack_status          | CREATE_FAILED
| stack_status_reason   | Resource CREATE failed: ResourceInError:
|                       | resources.Server1: Went to status ERROR due to
|                       | "Message: Exceeded maximum number of retries. Exceeded
|                       | max scheduling attempts 3 for instance
|                       | 2048d06f-d030-4ce1-99f5-3525eb4efd45. Last exception:
|                       | Binding failed for port 3c3a55e3-a25a-
|                       | 44a2-9900-59ea6cd423b8, please check neutron logs for
|                       | more information., Code: 500"
| stack_user_project_id | 41ed7a413243443ca49c7779fa56a74f
| tags                  | null
| template_description  | Launch a basic instance with centos7.2 image using the
|                       | ``m1.small`` flavor and one network.
| timeout_mins          | None
| updated_time          | None
```

图 8 -9　查看栈详细信息

(5) 查看栈资源列表，执行命令如下：

```
[root@ controller ~] # heat resource - list orchestration
```

执行返回结果如图 8 -10 所示。

```
[root@controller ~]# heat resource-list orchestration
+---------------+--------------------------------------+------------------+-----------------+---------------------+
| resource_name | physical_resource_id                 | resource_type    | resource_status | updated_time        |
+---------------+--------------------------------------+------------------+-----------------+---------------------+
| Server1       | 1934a430-a32d-4e7e-8e9e-aea269769a0d | OS::Nova::Server | CREATE_FAILED   | 2019-11-27T02:46:36 |
+---------------+--------------------------------------+------------------+-----------------+---------------------+
```

图 8－10 查看栈资源列表

（6）查看栈资源，执行命令如下：

[root@ controller ~] # heat resource - show orchestration server1

执行返回结果如图 8－11 所示。

```
[root@controller ~]# heat resource-show orchestration Server1
+------------------------+---------------------------------------------------------------------------------------
| Property               | Value
+------------------------+---------------------------------------------------------------------------------------
| attributes             | {
|                        |   "OS-EXT-STS:task_state": null,
|                        |   "addresses": {},
|                        |   "links": [
|                        |     {
|                        |       "href": "http://controller:8774/v2.1/d2b88800df664a8cb196ec75034c9e45/servers/1934a
|                        |       "rel": "self"
|                        |     },
|                        |     {
|                        |       "href": "http://controller:8774/d2b88800df664a8cb196ec75034c9e45/servers/1934a430-a
|                        |       "rel": "bookmark"
|                        |     }
|                        |   ],
```

图 8－11 查看栈资源

（7）查看栈输出列表，执行命令如下：

[root@ controller ~] # heat output - list orchestration

执行返回结果如图 8－12 所示。

```
[root@controller ~]# heat output-list orchestration
+---------------+-----------------------------+
| output_key    | description                 |
+---------------+-----------------------------+
| instance_ip   | IP address of the instance. |
| instance_name | Name of the instance.       |
+---------------+-----------------------------+
```

图 8－12 查看栈输出列表

（8）查看栈输出值，执行命令如下：

```
[root@ controller ~] # heat output - show orchestration instance_ip
```

执行返回结果如图 8 - 13 所示。

```
[root@controller ~]# heat output-show orchestration instance_ip
"10.0.0.102"
```

图 8 - 13　查看栈输出值

(9) 查看栈事件列表，执行命令如下：

```
[root@ controller ~] # heat event - list orchestration
```

执行返回结果如图 8 - 14 所示。

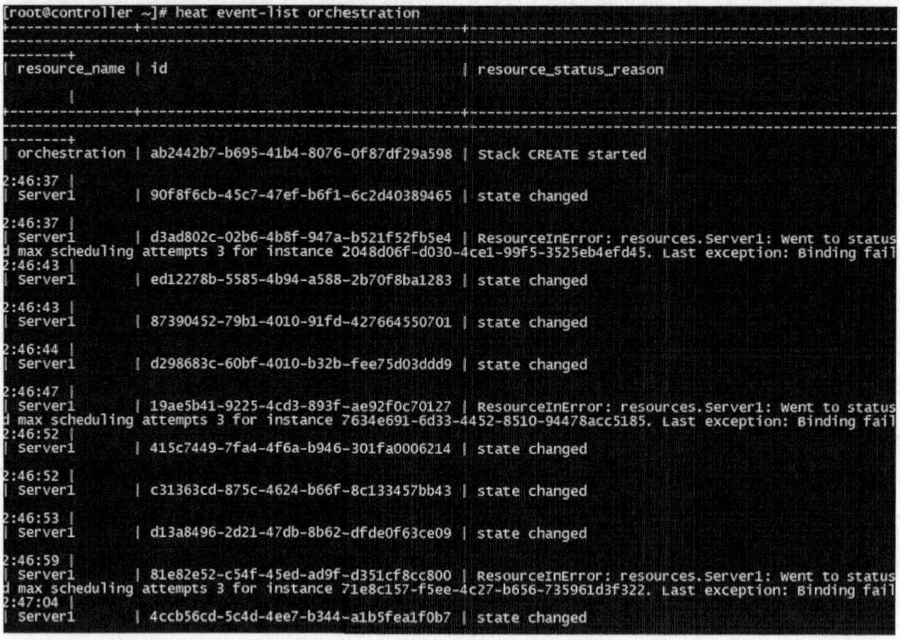

```
[root@controller ~]# heat event-list orchestration
+---------------+--------------------------------------+------------------------------------
--------------------------------------------------------------------------------------------
-------+
| resource_name | id                                   | resource_status_reason
       |
+---------------+--------------------------------------+------------------------------------
--------------------------------------------------------------------------------------------
-------+
| orchestration | ab2442b7-b695-41b4-8076-0f87df29a598 | Stack CREATE started
2:46:37 |
| Server1       | 90f8f6cb-45c7-47ef-b6f1-6c2d40389465 | state changed
2:46:37 |
| Server1       | d3ad802c-02b6-4b8f-947a-b521f52fb5e4 | ResourceInError: resources.Server1: Went to status
d max scheduling attempts 3 for instance 2048d06f-d030-4ce1-99f5-3525eb4efd45. Last exception: Binding fail
2:46:43 |
| Server1       | ed12278b-5585-4b94-a588-2b70f8ba1283 | state changed
2:46:43 |
| Server1       | 87390452-79b1-4010-91fd-427664550701 | state changed
2:46:44 |
| Server1       | d298683c-60bf-4010-b32b-fee75d03ddd9 | state changed
2:46:47 |
| Server1       | 19ae5b41-9225-4cd3-893f-ae92f0c70127 | ResourceInError: resources.Server1: Went to status
d max scheduling attempts 3 for instance 7634e691-6d33-4452-8510-94478acc5185. Last exception: Binding fail
2:46:52 |
| Server1       | 415c7449-7fa4-4f6a-b946-301fa0006214 | state changed
2:46:52 |
| Server1       | c31363cd-875c-4624-b66f-8c133457bb43 | state changed
2:46:53 |
| Server1       | d13a8496-2d21-47db-8b62-dfde0f63ce09 | state changed
2:46:59 |
| Server1       | 81e82e52-c54f-45ed-ad9f-d351cf8cc800 | ResourceInError: resources.Server1: Went to status
d max scheduling attempts 3 for instance 71e8c157-f5ee-4c27-b656-735961d3f322. Last exception: Binding fail
2:47:04 |
| Server1       | 4ccb56cd-5c4d-4ee7-b344-a1b5fea1f0b7 | state changed
```

图 8 - 14　查看栈事件列表

(10) 查看栈资源事件详细信息，执行命令如下：

```
[root@ controller ~] # heat event - show orchestration Server1 06fbeab4 -3f50 -4025 -8ed0 -d4a81db070e0
```

执行返回结果如图 8 - 15 所示。

(11) 删除栈，执行命令如下：

```
[root@ controller ~] # heat stack - delete orchestration
```

执行返回结果如图 8 - 16 所示。

```
[root@controller ~]# heat event-show orchestration Server1  90f8f6cb-45c7-47ef-b6f1-6c2d40389465
+------------------------+--------------------------------------------------------------------------------
| Property               | Value
+------------------------+--------------------------------------------------------------------------------
| event_time             | 2019-11-27T02:46:37
| id                     | 90f8f6cb-45c7-47ef-b6f1-6c2d40389465
| links                  | http://controller:8004/v1/d2b88800df664a8cb196ec75034c9e45/stacks/orchestration/470ea081-f5
6cb-45c7-47ef-b6f1-6c2d40389465 (self)
|                        | http://controller:8004/v1/d2b88800df664a8cb196ec75034c9e45/stacks/orchestration/470ea081-f5
|                        | http://controller:8004/v1/d2b88800df664a8cb196ec75034c9e45/stacks/orchestration/470ea081-f5
| logical_resource_id    | Server1
| physical_resource_id   | None
| resource_name          | Server1
| resource_properties    | {
|                        |   "admin_pass": null,
|                        |   "user_data_update_policy": "REPLACE",
|                        |   "availability_zone": null,
|                        |   "personality": {},
|                        |   "image": "e9a5f197-ccef-4f9b-8015-f6b1881e7c71",
|                        |   "user_data": "",
|                        |   "diskConfig": null,
|                        |   "flavor_update_policy": "RESIZE",
|                        |   "flavor": "2",
|                        |   "reservation_id": null,
```

图 8－15　查看栈资源事件详细信息

```
[root@controller ~]# heat stack-delete orchestration
Are you sure you want to delete this stack(s) [y/N]? y
+--------------------------------------+---------------+---------------+----------------------+--------------+
| id                                   | stack_name    | stack_status  | creation_time        | updated_time |
+--------------------------------------+---------------+---------------+----------------------+--------------+
| 470ea081-f598-4f47-9ed5-b6dbbe35c92d | orchestration | CREATE_FAILED | 2019-11-27T02:46:36  | None         |
+--------------------------------------+---------------+---------------+----------------------+--------------+
```

图 8－16　删除栈

5. 创建符合企业要求的编配任务

（1）编写 RD_Dept. yml 文件。编辑 RD_Dept. yml 文件内容，执行命令如下：

```
[root@ controller ~] # vi RD_Dept.yml
```

按【I】键进入编译模式，输入以下内容：

```
heat_ template_ version: 2015 -10 -15
 description: Launch a basic instance with Centos7.2  image
using the //m1 . small //flavor and one network.
parameters:
  Net:
    type: string
    description: Network for the instance.
resources:
  Server1:
    type: OS:: Nova:: Server
    properties:
      image: Centos6.5
      flavor: m1 . small
```

```
      networks:
       - network: { get_ param: Net }
  Server2:
    type: OS:: Nova:: Server
    properties:
      image: Centos7.2
      flavor: m1. small
      networks:
       - network: { get_ param: Net }
outputs:
Server1_ instance_ name:
  description: Name of the instance.
  value: { get_ attr: [ Server1, name ] }
Server1_ instance_ ip:
  description: IP address of the instance.
  value: { get_ attr: [ Server1, first_ address ] }
Server2_ instance_ name:
  description: Name of the instance.
  value: { get_ attr: [ Server2, name ] }
Server2_ instance_ ip:
  description: IP address of the instance.
  value: { get_ attr: [ Server2, first_ address ] }
```

按【Esc】键并输入：wq 命令保存退出。

（2）通过 Dashboard 界面为项目研发部启动栈资源，创建实例。

①通过 Dashboard 界面“项目”选项，打开编配面板，找到“栈”，选择“启动栈”选项，在弹出窗口中输入模板源和环境源，如图 8－17 所示。

图 8－17　为实例选择模板

②输入栈名、密码和网络，单击“运行”按钮，如图 8－18 所示。

图 8－18　启动栈

③创建成功后，查看栈的信息，如图 8－19 所示。

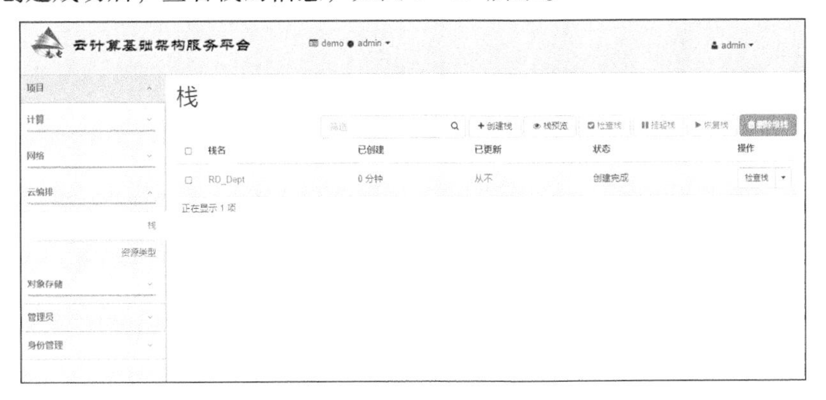

图 8－19　查看栈信息

④查看创建栈的实例，如图 8－20 所示。

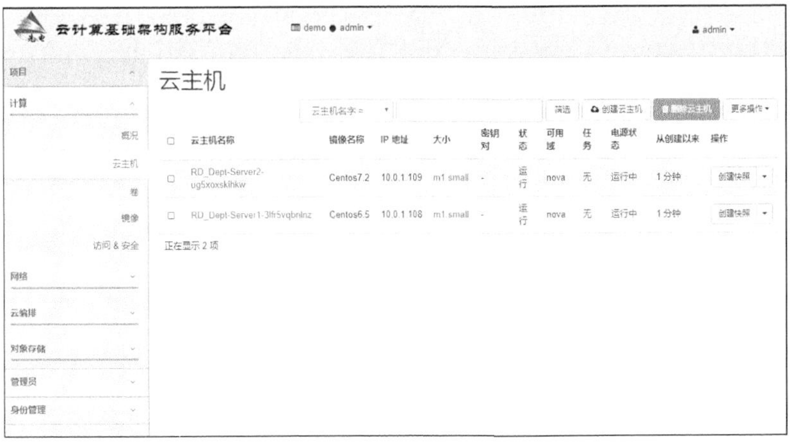

图 8－20　查看创建栈的实例

(3) 通过 CLI 命令行为业务部启动栈资源，创建实例。

```
[root@ controller ~] # heat stack-create -f RD_Dept.yml
--parameters " Net = BS_ Dept" BS_Dept
```

(4) 通过 CLI 命令行为为 IT 工程部启动栈资源，创建实例。

```
[root@ controller ~] # heat stack-create -f RD_Dept.yml
--parameters " Net = IT_ Dept" IT_Dept
```

任务二　安装与操作云监控服务

任务要求

本任务中小张要学习的是监控服务，通过监控服务可以实时查看平台的运行情况，保障平台运行稳定，维护数据安全，对可能出现的危险做到快速判断和处理。要求可以通过 Dashboard 界面和 CLI 查看平台某一个时间段的运行数据，包括网络数据、实例数据、存储数据和服务资源消耗情况。

核心概念

云监控服务：OpenStack 云监测服务 Ceilometer 承担着云服务的“计费”的功能，为上层的计费、结算或者监控应用使用数据收集功能提供统一的资源。

知识准备

监控服务的相关知识

1. 基本概念

(1) Ceilometer 的意义通常来说，云服务尤其是公有云服务，除了提供基本服务外，还承担了“计费”的功能，公有云在计费方面有 3 个层次。

① 计量（Metering）：收集资源的使用数据，其数据信息主要包括使用对象、使用者、使用时间和用量。

② 计费（Rating）：将资源使用数据按照商务规则转化为可计费项目并计算费用。

③ 结算（Billing）：收钱开票。Ceilometer 的目标是计量方面，为上层的计费、结算或者监控应用使用数据收集功能提供统一的资源。

(2) Ceilometer 的主要概念包括以下 5 类。

① Meter：是资源使用的计量项，它的属性包括名称（Name）、单位（Unit）、类型（Cumulative：累计值；Delta：变化值；Gauge：离散或者波动值）以及对应的资源属性等。

② Sample：是某个时刻某个资源（Resource）的某个 Meter 值。Sample 的集有

区间概念，即收集数据的时间间隔。除了 Meter 属性外，还有 Timestampe（采样时间）和 Volume（采样值）属性。

③ Statistics：一个时间段（Period）内的 Samples 聚合值，包括计数（Count）、最大（Max）、最小（Min）、平均（Avg）、求和（Sum）等。

④ Resource：指被监控的资源对象，可以是一台虚拟机、一台物理机或者一个云硬盘。

⑤ Alarm：是 Ceilometer 的告警机制，可以通过阈值或者组合条件告警，并设置告警时触发的 action。

2. Meter 的数据处理

Meters 数据的处理使用 Pipeline 的方式，即 Meters 数据依次经过（零个或者多个）Transformer 和（一个或者多个）Publisher 处理，最后达到（一个或者多个）Receiver。其中，Receivers 包括 Ceilometer Collector 和外部系统。

3. Publisher 分发器

表 8－3 说明了 Ceilometer 支持的 Publisher 类型。

表 8－3　Publisher 类型说明

Publishers 类型	格　式	说　明	配　置　项
Notifier	notifier//? option 1 = value 1&option 2 = value2	Samples 数据被发到 AMQP 系统，然后被 Ceilometer collecter 接收。默认的 AMQP Queue 是 metering_topic = metering。这默认的方式	[publisher_notifier] metering_driver = messagingv2 metering_topic = metering
RPC	rpc：//? option 1 = value 1&option2 = value2	与 notufuer 类似，同样经过 AMQP，不过是同步操作，因此可能有性能问题	[publisher_rpc] metering_topic = metering
UDP	udp：// <host>：<port>/	Jingguo UDP port 发出。默认的 UDP 端口是 4952	Udp_port = 4952
File	file：pate? option 1 = value 1&option2 = value 1	发送到文件保存	

4. 数据保存

（1）Ceilometer Collector 从 AMQP 接收到数据后，会原封不动地通过一个或者多个分发器（Dispatchers）将它保存到指定位置。

支持的分发器如下：

① 文件分发器：保存到文件。通过添加配置项 dispatcher = file 来指定分发器类型为文件。

② HTTP 分发器：保存到外部的 HTTP target。通过添加配置项 dispatcher = http 来指定分发器类型为 HTTP。

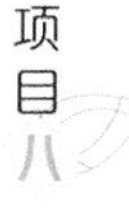

③ 数据库分发器：保存到数据库。添加配置项 dispatcher = database 来指定分发器类型为数据库。

（2）Ceilometer Collector 支持的数据库类型分发器有以下 3 种，如图 8－21 所示。

① MongoDB Database：默认 DB。

② SQL Database：支持 MySQL、PostgreSQL 和 IBM DB2 等。

③ HBase Database：DB。

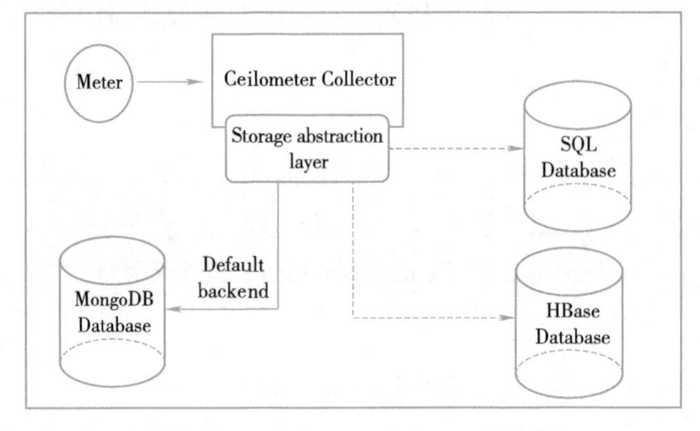

图 8－21　Ceilometer Collector 架构图

任务实施

监控服务的基本操作

1. 准备实施任务的环境（见表 8－4）

表 8－4　实施任务的准备环境

实施任务所需软件资源	虚拟机镜像资源信息
虚拟机软件（vm12）	controller
securcrt 远程连接软件	compute

2. 分解实施任务的过程（见表 8－5）

表 8－5　实施任务的简明过程

序号	步骤	详细操作及说明
1	云监控服务的基本操作	①安装云监控服务； ②数据查看； ③数据库备份
2	使用 Dashboard 界面上进行数据查看	①查看资源使用情况； ②通过 Dashboard 或 CLI 查看平台某一个时间段的运行数据

3. 云监控服务 Ceilometer 的操作

（1）在控制和计算节点上分别安装 Ceilometer 组件，如图 8－22 所示。

```
[root@ controller ~] #source /etc/keystone/admin-openrc.sh
[root@ controller ~] #vi /etc/xiandian/openrc.sh
```

按【I】键进入编译模式，将文件修改为以下形式：

```
CEILOMETER_ DBPASS =000000
CEILOMETER_ PASS =000000
```

按【Esc】键并输入：wq 命令保存退出。

安装 Ceilometer 组件：

```
[root@ controller ~] # iaas - install - ceilometer -
controller. sh
[root@ compute ~] # vi /etc/xiandian/openrc. sh
[root@ compute ~] # source /etc/keystone/admin - openrc. sh
[root@ compute ~] # vi /etc/xiandian/openrc. sh
```

按【I】键进入编译模式，将文件修改为以下形式：

```
CEILOMETER_ DBPASS =000000
CEILOMETER_ PASS =000000
```

按【Esc】键并输入：wq 命令保存退出。

安装 Ceilometer 组件：

```
[root@ compute ~] # iaas - install - ceilometer - compute. sh
```

```
iaas-install-trove.sh
iaas-pre-host.sh
iaas-uninstall-all.sh
[root@compute bin]# iaas-install-ceilometer-compute.sh
```

图 8－22　安装 Ceilometer 组件

（2）数据查看。

可以用命令查看网络数据、实例数据、存储数据和服务资源消耗情况。

在 OpenStack 系统中进行操作需生效环境变量，执行命令如下：

```
[root@ controller ~] # source /etc/keystone/admin - openrc.sh
```

查看网络数据，执行命令如下：

```
[root@ controller ~] # ceilometer statistics -m network.
incoming. bytes
```

执行返回结果如图 8－23 所示。

```
[root@controller ~]# ceilometer statistics -m network.incoming.bytes
+---------+----------------------------+----------------------------+---------+---------+-
-------+---------+-------+----------+----------------------------+--------------------
------+
| Period | Period Start                | Period End                 | Max     | Min     |
Avg     | Sum     | Count | Duration | Duration Start             | Duration End
      |
+---------+----------------------------+----------------------------+---------+---------+-
-------+---------+-------+----------+----------------------------+--------------------
------+
| 0      | 2019-03-15T10:26:52.039000 | 2019-03-16T16:36:25.343000 | 1612.0 | 1612.0 |
1612.0 | 1612.0 | 1     | 0.0      | 2019-03-16T16:36:25.343000 | 2019-03-16T16:36:25.3
43000 |
+---------+----------------------------+----------------------------+---------+---------+-
-------+---------+-------+----------+----------------------------+--------------------
```

图 8－23　查看网络数据

查看实例数据，执行命令如下：

```
[root@ controller ~] # ceilometer statistics -m instance
```

执行返回结果如图 8-24 所示。

```
[root@controller ~]# ceilometer statistics -m instance
+--------+--------------------------------+--------------------------------+-----+-----+-----+-
---+-------+--------------------+--------------------------------+
| Period | Period Start                   | Period End                     | Max | Min | Avg |
Sum | Count | Duration | Duration Start                 | Duration End                   |
+--------+--------------------------------+--------------------------------+-----+-----+-----+-
---+-------+--------------------+--------------------------------+
| 0      | 2019-03-15T10:26:52.039000 | 2019-03-16T16:36:25.388000 | 1.0 | 1.0 | 1.0 |
1.0 | 1     | 0.0        | 2019-03-16T16:36:25.388000 | 2019-03-16T16:36:25.388000 |
+--------+--------------------------------+--------------------------------+-----+-----+-----+-
```

图 8-24　查看实例数据

查看存储数据，执行命令如下：

```
[root@ controller ~] # ceilometer statistics -m disk.read.
requests
```


执行返回结果如图 8-25 所示。

```
[root@controller ~]# ceilometer statistics -m disk.read.requests
+--------+--------------------------------+--------------------------------+--------+--------+-
-------+--------+-------+----------+--------------------------------+-----------------------
------+
| Period | Period Start                   | Period End                     | Max    | Min    |
Avg    | Sum    | Count | Duration | Duration Start                 | Duration End
       |
+--------+--------------------------------+--------------------------------+--------+--------+-
-------+--------+-------+----------+--------------------------------+-----------------------
------+
| 0      | 2019-03-15T10:26:52.039000 | 2019-03-16T16:36:25.452000 | 1099.0 | 1099.0 |
1099.0 | 1099.0 | 1     | 0.0      | 2019-03-16T16:36:25.452000 | 2019-03-16T16:36:25.4
52000 |
+--------+--------------------------------+--------------------------------+--------+--------+-
-------+--------+-------+----------+--------------------------------+-----------------------
```

图 8-25　查看存储数据

查看服务资源消耗情况，执行命令如下：

```
[root@ controller ~] # ceilometer statistics -m memory
```

执行返回结果如图 8-26 所示。

```
[root@controller ~]# ceilometer statistics -m memory
+--------+--------------------------------+--------------------------------+-------+-------+---
---+-------+-------+----------+--------------------------------+--------------------------------
--+
| Period | Period Start                   | Period End                     | Max   | Min   | Av
g    | Sum   | Count | Duration | Duration Start                 | Duration End
  |
+--------+--------------------------------+--------------------------------+-------+-------+---
---+-------+-------+----------+--------------------------------+--------------------------------
--+
| 0      | 2019-03-15T10:26:52.039000 | 2019-03-16T16:27:24.553000 | 512.0 | 512.0 | 51
2.0 | 512.0 | 1     | 0.0      | 2019-03-16T16:27:24.553000 | 2019-03-16T16:27:24.55300
0 |
+--------+--------------------------------+--------------------------------+-------+-------+---
---+-------+-------+----------+--------------------------------+--------------------------------
```

图 8-26　查看服务资源消耗情况

（3）数据库备份。

辅助 Shell 脚本做到对数据库的全量备份和增量备份，保证数据安全。

① 编写全量备份脚本内容，执行命令如下：

```
[root@ controller ~] # vi  mysql_full_bk.sh
```

按【I】键进入编译模式，输入以下内容：

```
#! /bin/bash
today ='date  + "% Y_% m_% d"'
bkdir = /opt /mysql /backup /
full_bkdir = $ bkdir /full_backup
today_bkdir = $ full_bkdir /$ today
DB_Name = root
DB_PASS = 000000
if [ ! -d $ bkdir ] ; then
    mkdir -p $ bkdir
else
    chattr -i $ bkdir
fi
if [ ! -d $ full_bkdir ] ; then
mkdir -p $ full_bkdir
else
    chattr -i $ full_bkdir
fi
if [ ! -d $ today_bkdir ] ; then
    mkdir -p $ today_bkdir
else
    chattr -i $ today_bkdir
fi
database = NOT_NULL
for ((i =2; i <10; i + +))
do
   if [ -n " $database" ]; then
   database = mysql  -u $ DB_Name  -p $ DB_PASS  -S /var/lib/
mysql /mysql. sock -e " show databases;"  | sed '1,' $ i 'd' | sed
2, $ d
      mysqladmin  -u $ DB_Name  -p $ DB_PASS flush -tables
      mysqldump  -u $ DB_Name  -p $ DB_PASS  $ database  -A -
- events  ignore  - tables  = mysql. events  >  $ today _ bkdir/
$ database. sql
      cd  $ full_bkdir
      for d in "  find.  -type d -mtime +6  -maxdepth 1"  > /
dev /null 2 >&1
      do
         chattr  -i  $ d
```

```
        rm -fr $d
      done
      chattr +i $bkdir
      echo " $database is fine"
      else
      break
      fi
done
```

按【Esc】键并输入：wq 命令保存退出。

将 mysql_full_bk. sh 脚本赋予执行权限，执行命令如下：

```
[root@ controller ~] # chmod +x mysql_full_bk.sh
```

② 开启 MySQL 数据库的 binlog 功能。

编辑/etc/my. cnf 文件，执行命令如下：

```
[root@ controller ~] # vi /etc/my.cnf
```

按【I】键进入编译模式，在［mysqld_safe］标签下添加 log - bin = mysql - bin 参数。具体如图 8 - 27 所示。

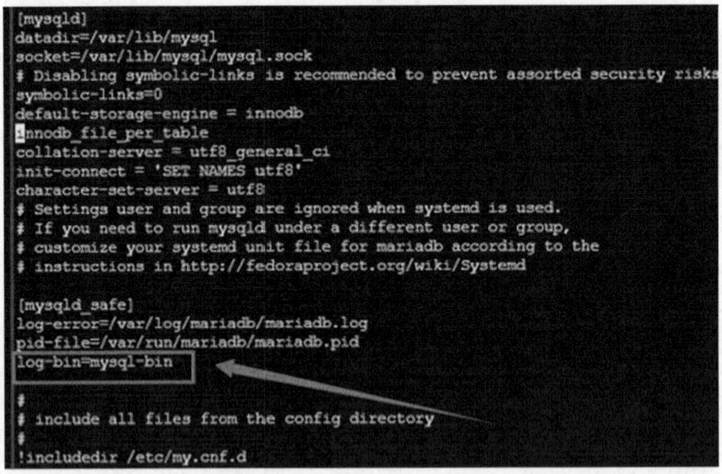

```
[mysqld]
datadir=/var/lib/mysql
socket=/var/lib/mysql/mysql.sock
# Disabling symbolic-links is recommended to prevent assorted security risks
symbolic-links=0
default-storage-engine = innodb
innodb_file_per_table
collation-server = utf8_general_ci
init-connect = 'SET NAMES utf8'
character-set-server = utf8
# Settings user and group are ignored when systemd is used.
# If you need to run mysqld under a different user or group,
# customize your systemd unit file for mariadb according to the
# instructions in http://fedoraproject.org/wiki/Systemd

[mysqld_safe]
log-error=/var/log/mariadb/mariadb.log
pid-file=/var/run/mariadb/mariadb.pid
log-bin=mysql-bin

#
# include all files from the config directory
#
!includedir /etc/my.cnf.d
```

图 8 - 27　/etc/my. cnf 文件内容

按【Esc】键并输入：wq 命令保存退出。

重启数据库服务，执行命令如下：

```
[root@ controller ~] # systemctl restart mariadb
```

编写增量备份脚本 mysql_hourly_bk. sh。执行命令如下：

```
[root@ controller ~] # vi  mysql_hourly_bk.sh
```

按【I】键进入编译模式，输入以下内容：

```
#! /bin/bash
today=date +"%Y_%m_%d"
bkdir=/opt/mysql/backup/
hourly_bkdir=$bkdir/hourly_backup
today_bkdir=$hourly_bkdir/$today
DB_Name=root
DB_PASS=000000
log_dir=/var/lib/mysql
if [ ! -d $bkdir ] ; then
     mkdir -p $bkdir
else
     chattr -i $bkdir
fi
if [ ! -d $hourly_bkdir ] ; then
     mkdir -p $hourly_bkdir
else
     chattr -i $hourly_bkdir
fi
if [ ! -d $today_bkdir ] ; then
     mkdir -p $today_bkdir
else
     chattr -i $today_bkdir
fi
mysqladmin -u$DB_Name -p$DB_PASS flush-logs
total=ls $log_dir/mysql-bin.* |wc -l
total=expr $total - 2
for f in ls $log_dir/mysql-bin.* |head -n $total
do
     bf=basename $f
     echo $bf is finish.
     mv $f $today_bkdir
done
```

按【Esc】键并输入:wq 命令保存退出。

将 mysql_hourly_bk.sh 赋予执行权限，执行命令如下：

```
[root@ controller ~]# chmod +x mysql_hourly_bk.sh
```

③ 开启定时设置，编写定时任务内容，执行命令如下：

```
[root@controller ~]#crontab -e
```

按【I】键进入编译模式，输入以下内容：

```
0 1-23/3 * * * (/bin/sh /root/mysql_hourly_bk.sh >>/opt/mysql/backup/backup.log)    //在1至23时内每个3小时执行全量备份命令
0 0 * * 0,1,3,6 (/bin/sh /root/mysql_full_bk.sh >>/opt/mysql/backup/backup.log)    //每周一、三、六、日的00:00时执行增量备份命令
```

按【Esc】键并输入：wq命令保存退出。

重启定时服务，执行命令如下：

```
[root@controller ~]#systemctl restart crond
```

查看定时任务文件内容，执行命令如下：

```
[root@controller ~]#crontab -l
```

执行返回结果如图8-28所示。

```
[root@controller ~]# crontab -l
39 * * * *  (/bin/sh /root/mysql_hourly_bk.sh >> /opt/mysql/backup/backup.log)
0 0 * * 0,1,3,6 (/bin/sh /root/mysql_full_bk.sh >> /opt/mysql/backup/backup.log)
```

图8-28　定时任务文件内容

视频

完成监控任务

4. 使用Dashboard界面进行数据查看

（1）查看资源使用情况，如图8-29所示。

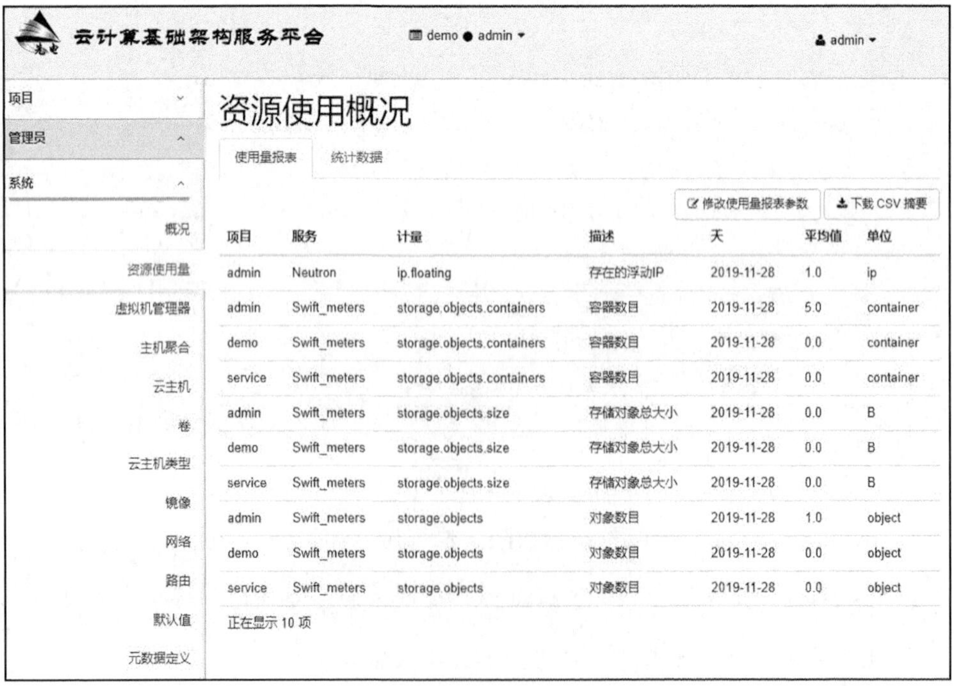

云计算基础架构服务平台　demo ● admin ▾　admin ▾

项目

管理员

系统

概况

资源使用量

虚拟机管理器

主机聚合

云主机

卷

云主机类型

镜像

网络

路由

默认值

元数据定义

资源使用概况

使用量报表　统计数据

修改使用量报表参数　下载CSV摘要

项目	服务	计量	描述	天	平均值	单位
admin	Neutron	ip.floating	存在的浮动IP	2019-11-28	1.0	ip
admin	Swift_meters	storage.objects.containers	容器数目	2019-11-28	5.0	container
demo	Swift_meters	storage.objects.containers	容器数目	2019-11-28	0.0	container
service	Swift_meters	storage.objects.containers	容器数目	2019-11-28	0.0	container
admin	Swift_meters	storage.objects.size	存储对象总大小	2019-11-28	0.0	B
demo	Swift_meters	storage.objects.size	存储对象总大小	2019-11-28	0.0	B
service	Swift_meters	storage.objects.size	存储对象总大小	2019-11-28	0.0	B
admin	Swift_meters	storage.objects	对象数目	2019-11-28	1.0	object
demo	Swift_meters	storage.objects	对象数目	2019-11-28	0.0	object
service	Swift_meters	storage.objects	对象数目	2019-11-28	0.0	object

正在显示10项

图8-29　查看资源使用情况

(2) 通过 Dashboard 或 CLI 查看平台某一个时间段的运行数据。

①使用 CLI 命令行查看网络数据、实例数据、存储数据及服务资源消耗情况：

```
[root@ controller ~] # ceilometer statistics -m network.
incoming.bytes
```

②查看实例数据：

```
[root@ controller ~] #ceilometer statistics -m instance
```

③查看存储数据：

```
[root@ controller ~] #ceilometer statistics -m disk.read.
requests
```

④查看服务资源消耗情况：

```
[root@ controller ~] #ceilometer statistics -m memory
```

本项目所讲解的 Heat 和 Ceilometer 组件是 OpenStack 核心组件之外的组件，主要是提高平台快速部署实例及提升平台的监控管理能力。主要学习 Heat 模板的编写方法及利用 Heat 模板实现实例的批量自动部署和利用监控服务实时查看云平台的运行情况。

表 8-6 列举本项目 Heat 和 Ceilometer 的常用命令及其说明。

表 8-6　常用命令及其说明

常用命令	命令说明
heat action-resume < NAME or ID >	重启或恢复栈
heat action-suspend < NAME or ID >	挂起栈
heat build-info	获取建栈的信息
heat event-list [-r < RESOURCE >] < NAME or ID >	stack 的事件列表
heat event-show < NAME or ID > < RESOURCE > < EVENT >	描述 stack 的事件
heat output-list < NAME or ID >	显示可用的输出
heat output-show < NAME or ID > < OUTPUT NAME >	显示可用输出的值
heat resource-list < NAME or ID >	显示某个 stack 资源的列表
heat stack-create	创建一个栈
heat stack-adopt	使用栈
heat stack-delete	删除一个栈

续表

常用命令	命令说明
heat stack - list	列出用户所有的栈
ceilometer meter - list	查询现在所有监控的资源
ceilometer sample - list - m cpu	查询某种监控资源
ceilometer statistics - - meter cpu_ util	查询某种资源的统计信息
ceilometer alarm - list	查询现在所有的 alarm

结合本项目的 Heat 组件的理念和命令，思考要创建一个能实现两台运行 Apache 的 Web 服务器的栈，应该如何做？

项目九

综合实训

项目综述

小张在学习完搭建云平台及云平台的相关组件后，需能按照企业的应用需求搭建私有云平台，实现云计算平台架构的规划设计，完成云计算平台网络交换机设备、服务器、存储服务器的互联和配置且云计算基础架构平台部署、配置和管理，满足应用场景需求。小张为此要完成的任务如下：

项目九综述

- 云计算平台架构的规划设计。
- 完成云计算平台网络交换机设备、服务器、存储服务器的互联和配置。
- 完成云计算基础架构平台部署、配置和管理。

项目目标

【知识目标】

- OpenStack 的各服务组件的相关概念。
- OpenStack 的各服务流程和工作机制。

【技能目标】

- 云计算平台架构的规划设计。
- 完成云计算平台网络基础设备、服务器、存储服务器的互联和配置。
- 完成云计算基础架构平台部署、配置和管理。

【职业能力】

根据企业的应用需求，完成私有云平台搭建，满足应用场景需求。

任务　搭建 IaaS 平台系统

任务要求

某企业计划搭建私有云平台，以实现计算资源的池化弹性管理，企业应用的集中管理，统一安全认证和授权管理。需完成云平台架构的设计、系统部署、系统管理。试根据用户需求，完成以下任务。

1. 准备实施任务的环境（见表 9－1）

表 9－1 实施任务的准备环境

实施任务所需软件资源	虚拟机镜像资源信息
虚拟机软件（vm12）	CentOS－7－x86_64－DVD－1511. iso
securcrt 远程连接软件	Xiandian－IaaS－2. 2. iso

2. 分解实施任务的过程（见表 9－2）

表 9－2 实施任务的简明过程

序号	步　骤	详细操作及说明
1	根据企业需求设计云平台的架构	
2	环境配置	①使用相关命令查询控制节点和计算节点主机名； ②在控制节点和计算节点配置主机名映射，实现云平台管理网络地址与主机名的映射； ③各个节点关闭防火墙，设置开机不启动，设置 selinux 为 permissive，使用 getenforce 命令进行查询
3	yum 安装	①在控制及计算节点挂载镜像制作源路径； ②在控制及计算节点配置 yum 路径； ③在控制及计算节点创建 repo 文件
4	配置 IP 地址	根据规划配置控制及计算节点的 IP 地址
5	在控制节点和计算节点的虚拟机安装 iaas－xiandian 及 iaas－pre－host 软件包，并编辑文件/etc/xiandian/openrc. sh（配置环境变量）	根据规划配置配置环境变量
6	OpenStack 基础架构平台安装	参考项目一安装 OpenStack 软件的基础部分
7	数据库管理	参考数据库 MySQL 的常用命令
8	Keystone 管理	参考项目二的 Keystone 常用命令
9	Glance 管理	参考项目三的 Glance 常用命令
10	Nova 管理	参考项目六的 Nova 常用命令
11	Neutron 管理	参考项目五的 Neutron 常用命令

3. 根据企业需求设计云平台的架构

根据企业的需求，系统架构如图 9－1 所示，IP 地址规划如表 9－3 所示。

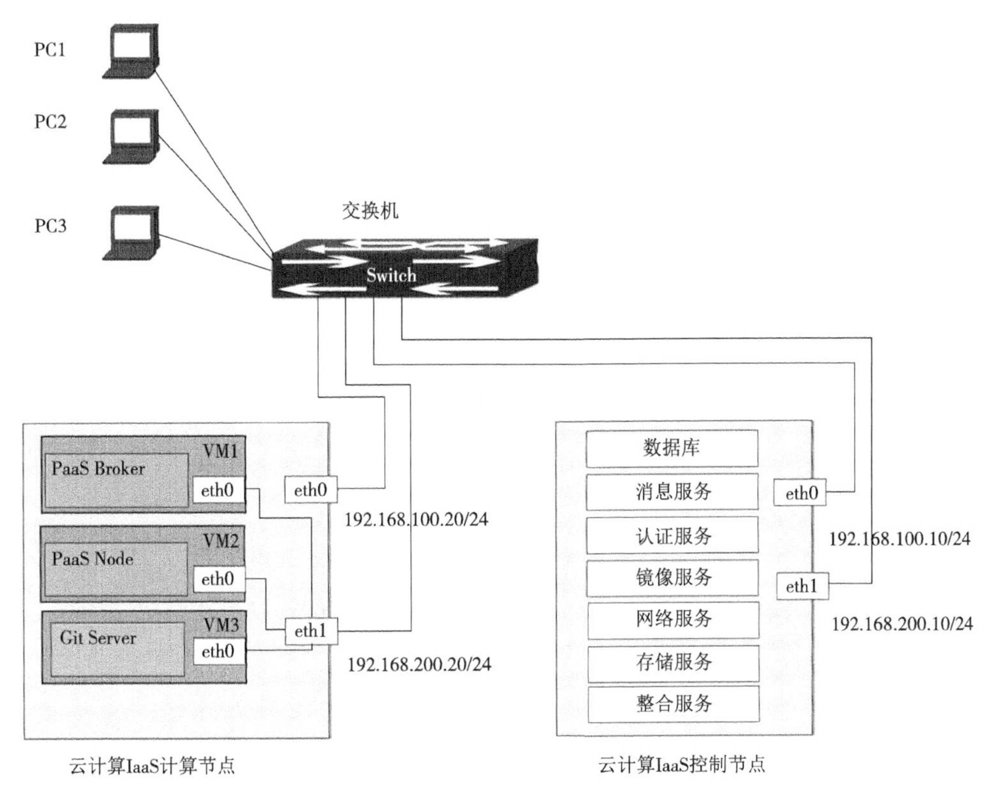

图 9－1　系统架构

表 9－3　IP 地址规划

设备名称	接口	IP 地址
控制节点	eth0	192. 168. 100. 10/24
	eth1	192. 168. 200. 10/24
计算节点	eth0	192. 168. 100. 20/24
	eth1	192. 168. 200. 20/24

4. 环境配置

手动配置云平台 IaaS 各节点的系统参数。

(1) 使用相关命令设置控制节点和计算节点主机名。

```
[root@ controller ~] # [root@ localhost ~] # hostnamectl set -
hostname controller
controller
[root@ compute ~] # [root@ localhost ~] # hostnamectl set
-hostname compute
compute
```

(2) 在控制节点和计算节点配置主机名映射，实现云平台管理网络地址与主机名的映射。

```
[root@ controller ~] # vi /etc/hosts
```

按【I】键进入编译模式，添加以下内容：

```
192.168.100.10 controller
[root@ compute ~] # vi /etc/hosts
```

按【I】键进入编译模式，添加以下内容：

```
192.168.100.20 compute
```

（3）各个节点关闭防火墙，设置开机不启动，设置 selinux 为 permissive，使用 getenforce 命令进行查询。

清除所有 chains 链（INPUT/ OUTPUT/ FORWARD）中所有的 rule 规则：

```
[root@ controller ~] # iptables -F
```

清空所有 chains 链（INPUT/ OUTPUT/ FORWARD）中包及字节计数器：

```
[root@ controller ~] # iptables -X
```


清除用户自定义 chains 链（INPUT/ OUTPUT/ FORWARD）中的 rule 规则：

```
[root@ controller ~] # iptables -Z
```

执行清除命令没有返回结果，通过/ usr/sbin/ iptables - save 保存：

```
[root@ controller ~] # /usr/sbin/iptables-save
```

清除所有 chains 链（INPUT/ OUTPUT/ FORWARD）中所有的 rule 规则：

```
[root@ compute ~] # iptables -F
```

清空所有 chains 链（INPUT/ OUTPUT/ FORWARD）中包及字节计数器：

```
[root@ compute ~] # iptables -X
```

清除用户自定义 chains 链（INPUT/ OUTPUT/ FORWARD）中的 rule 规则：

```
[root@ compute ~] # iptables -Z
```

执行清除命令没有返回结果，通过/ usr/ sbin/ iptables - save 保存：

```
[root@ compute ~] # /usr/sbin/iptables-save
[root@ controller ~] # getenforce
[root@ compute ~] # getenforce
```

5. yum 安装

（1）在控制节点挂载镜像制作源路径。

在控制节点虚拟机上的 CDROM 上选择 CentOS - 7 - x86_64 - DVD - 1511. iso，接着挂载：

```
[root@ controller ~] # mount -o loop /dev/cdrom /mnt/
[root@ controller ~] # mkdir /opt/centos
[root@ controller ~] # cp -rf /mnt/* /opt/centos/
[root@ controller ~] # umount /mnt/
```

在控制节点虚拟机上的 CDROM 上选择 XianDian – IaaS – v2. 2. iso，接着挂载：

```
[root@ controller ~] # mount -o loop /dev/cdrom /mnt/
[root@ controller ~] # cp -rf /mnt/* /opt/
[root@ controller ~] # umount /mnt/
```

结果如图 9 – 2 所示。

```
[root@controller ~]# mount -o loop /dev/cdrom /mnt
[root@controller ~]# mkdir /opt/centos
[root@controller ~]# cp -rf /mnt/* /opt/centos
[root@controller ~]# umount /mnt/
[root@controller ~]#
[root@controller ~]# mount -o loop /dev/cdrom /mnt
[root@controller ~]# cp -rf /mnt/* /opt/
[root@controller ~]# umount /mnt
[root@controller ~]#
```

图 9 – 2 挂载情况 1

（2）配置 yum 路径。

从 yum 源路径移除 yum 目录：

```
[root@ controller ~] # mv /etc/yum.repos.d/* /opt
```

执行没有返回结果。

（3）在控制节点创建 repo 文件。

在/etc/yum. repos. d 创建 local. repo 源文件：

```
[root@ controller ~] # vi /etc/yum.repos.d/local.repo
```

按【I】键进入编译模式，添加以下内容到 local. repo 文件里：

```
[centos]
name=centos
baseurl=file:///opt/centos
gpgcheck=0
enabled=1
[iaas]
name=iaas
baseurl=file:///opt/iaas-repo
gpgcheck=0
enabled=1
```

按【Esc】键并输入：wq 命令保存退出。

（4）在计算节点挂载镜像制作源路径。

在计算节点上的 CDROM 上选择 CentOS – 7 – x86_64 – DVD – 1511. iso，接着挂载：

```
[root@ compute ~] # mount -o loop /dev/cdrom/mnt/
[root@ compute ~] # mkdir /opt/centos
[root@ compute ~] # cp -rf /mnt/* /opt/centos/
[root@ compute ~] # umount /mnt/
```

在计算节点虚拟机上的 CDROM 上选择 XianDian – IaaS – v2. 2. iso，接着挂载：

```
[root@ compute ~] # mount -o loop /dev/cdrom/mnt/
[root@ compute ~] # cp -rf /mnt/ * /opt/
[root@ compute ~] # umount /mnt/
```

结果如图 9 – 3 所示。

```
[root@compute ~]# mount -o loop /dev/cdrom /mnt
[root@compute ~]# mkdir /opt/centos
[root@compute ~]# cp -rf /mnt/* /opt/centos/
[root@compute ~]# umount /mnt/
[root@compute ~]#
[root@compute ~]# mount -o loop /dev/cdrom  /mnt/
[root@compute ~]# cp -rf /mnt/* /opt/
[root@compute ~]# umount /mnt/
[root@compute ~]#
```

图 9 – 3　挂载情况 2

（5）配置 yum 路径。

将网络 yum 源路径移除 yum 目录：

```
[root@ compute ~] # mv /etc/yum.repos.d/ * /opt
```

执行没有返回结果。

（6）在计算节点创建 repo 文件。

在/ etc/ yum. repos. d 创建 local. repo 源文件：

```
[root@ compute ~] # vi /etc/yum.repos.d/local.repo
```

按【I】键进入编译模式，添加以下内容到 local. repo 文件里：

```
[centos]
name = centos
baseurl = file: ///opt/centos
gpgcheck = 0
enabled = 1
[iaas]
name = iaas
baseurl = file: ///opt/iaas - repo
gpgcheck = 0
enabled = 1
```

按【Esc】键并输入：wq 命令保存退出。

6. 配置 IP 地址

根据服务器自身的 IP 地址配置网卡信息，可以参考项目一，设置好后，使用 ip a命令查看。

```
[root@ controller ~] # ip a
[root@ computer ~] # ip a
```

7. 平台配置

在控制节点和计算节点的虚拟机安装 iaas - xiandian 及 iaas - pre - host 软件包，并编辑文件/etc/xiandian/openrc. sh（配置环境变量），具体修改参数如表 9 - 4 所示。

表 9 - 4　云平台配置信息

服　务	用　户	密　码
MySQL	root	000000
	Keystone	000000
	Glance	000000
	Nova	000000
	Neutron	000000
	Swift	000000
	Cinder	000000
	Heat	000000
	Ceilometer	000000
Keystone	Admin	000000
	Glance	000000
	Nova	000000
	Neutron	000000
	Swift	000000
	Cinder	000000
	Heat	000000
	Ceilometer	000000

```
######## Basic Environment ########
#          数据库用户密码          #
Mysql_ Admin_ Passwd =000000
#          管理员密码              #
Admin_ Passwd =000000
#          演示用户密码            #
Demo_User_Passwd =000000
#          演示数据库密码          #
Demo_DB_Passwd =000000
###################################
# ------------------------------------------------------
--------
```

```
#  Controller Node
#------------------------------------------------------------
---------
#  Controller Node Hostname
##          控制节点主机名          ##
Controller_Hostname = controller
#  Management Network of Controller Node IP
##        控制节点管理地址          ##
Controller_Mgmt_IPAddress = 192.168.100.10
#  External Network of Controller Node IP   #
##            外部地址              ##
Controller_External_IPAddress = 192.168.200.10
#------------------------------------------------------------
---------
#  Compute Node
#------------------------------------------------------------
---------
#  Compute Node Hostname
##        计算节点主机名            ##
Compute_Hostname = compute
#  Management Network of Compute Node IP
##        计算节点管理地址          ##
Compute_Mgmt_IPAddress = 192.168.100.20
#  External Network of Compute Node IP
##        计算节点外部地址          ##
Compute_External_IPAddress = 192.168.200.20
#  Cinder Disk Name eg (md126p3)
##        Cinder 存储磁盘分区名称 ##
Stroage_Cinder_Disk = sdb
#  Swift Disk Name eg (md126p4)
##        Swift 存储磁盘分区名称   ##
Stroage_Swift_Disk = sdc
```

8. OpenStack 基础架构平台安装

在控制节点和计算节点安装相应的脚本，具体参考项目一云平台的安装，使用 curl 命令查询网址 http：//controller/Dashboard 并查询结果。

```
[root@ controller  ~]# curl  -L http://192.168.100.10/dashboard
```

9. 数据库管理

（1）使用提供的数据库脚本安装数据库 MySQL。使用 root 用户登录，查询数据库列表信息。

```
[root@ controller ~] # mysql -uroot -p000000
mysql > show databases;
```

（2）在数据库中创建数据库 cloudcompute，创建用户 xiandian，密码为 chinaskills，并赋予 xiandian 只有使用数据库 cloudcompute 的权限。

```
[root@ controller ~]#mysql -uroot -pmysqlpass
```

10. Keystone 管理

（1）admin-openrc.sh 文件在/etc/keystone/目录下，查询 Keystone 相关命令查询用户列表。

```
[root@ controller ~]#openstack user list
```

（2）使用 Keystone 相关命令，查询 admin 用户信息。

```
[root@ controller ~]#openstack user show admin
```

（3）使用 Keystone 相关命令，查询认证服务详细信息。

```
[root@ controller ~]#openstack catalog show keystone
```

（4）在 Keystone 中创建用户 testuser，密码为 password，将该用户分配给 admin 租户，赋予用户 testuser 的 admin 权限，完成之后在云平台中禁用该用户，并查询用户 testuser 的状态。

```
[root@ controller ~]#openstack user create --password
password --domain demo testuser
[root@ controller ~]#openstack role add --user testuser --
project admin admin
[root@ controller ~]#openstack user set --disable testuser
```

11. Glance 管理

（1）使用 Glance 相关命令查询镜像，并查询信息。

```
[root@ controller ~]#glance image-list
```

（2）使用镜像文件 CentOS-7.2-x86_64_XD.qcow2 创建 glance 镜像 CentOS 7.2,格式为 qcow2,查询镜像列表。

```
[root@ controller ~]#glance image-create --name " centos7.
2" --disk-format qcow2 --container-format bare --progress <
/opt/iaas/images/CentOS_7.2_x86_64_XD.qcow2
[root@ controller ~]# glance image-show 836a899a-3d8a-
408f-a02f-d5c5ecd1f7fd
```

12. Nova 管理

（1）使用 Nova 相关命令查询 Nova 所有服务所在主机的列表信息。

```
[root@ controller ~]#nova host-list
```

（2）在使用 Nova 相关命令查询云主机类型的列表信息。

```
[root@ controller ~]#nova flavor-list
```

（3）使用 Nova 相关命令查询平台资源使用情况，将命令和结果以文本形式提交。

```
[root@ controller ~]#nova usage-list
```

（4）修改云平台中默认每个 tenant 的实例配额为 15 个，在答题框填入该命令，查询修改后的默认配额信息并以文本形式提交。

```
[root@ controller ~] # nova quota -class -update --instances 15 default
[root@ controller ~] # nova quota -defaults
```

13. Neutron 管理

完成网络创建之后，使用 neutron agent - list 命令查询 Neutron 各服务的状态。

```
[root@ controller ~] # neutron agent -list
```

项目小结

经过前 8 章的学习，已经具备了云计算平台架构的规划设计，云计算平台交换机设备、服务器、存储服务器的互联和配置和云计算基础架构平台部署、配置和管理的能力。项目九是综合实战，通过任务实施，使读者能根据企业的应用需求，完成搭建私有云平台满足应用场景需求的项目目标。

项目扩展

本项目搭建的云平台使用的系统是 Linux 的发行版 CentOS，如果改成 Ubuntu 发行版来搭建云平台，需从 Ubuntu Cloud Archive 软件包资源库安装 OpenStack，可以方便地在 Ubuntu 服务器上搭建新版 OpenStack。使用该方法还能为 OpenStack 的稳定运行和升级带来更高的可靠性，因为其始终可以和主文档保持一致。另外，使用 Ubuntu CLoud Archive 可以在 Ubuntu14.04 LTS 平台使用不同的 OpenStack 发布版本。

项目思考

本教程搭建的云平台使用的系统是 Linux 的发行版 CentOS，如果改成 Ubuntu 发行版来搭建云平台，哪些部分会有区别？